Lecture Notes in Computer

Commenced Publication in 1973
Founding and Former Series Editors:
Gerhard Goos, Juris Hartmanis, and Jan van l

Mike Burmester   Alec Yasinsac (Eds.)

Secure Mobile Ad-hoc Networks and Sensors

First International Workshop, MADNES 2005
Singapore, September 20-22, 2005
Revised Selected Papers

 Springer

Volume Editors

Mike Burmester
Alec Yasinsac
Florida State University
Department of Computer Science
Tallahasee, Florida, FL 32306-4530, USA
E-mail: {burmester,yasinsac}@cs.fsu.edu

Library of Congress Control Number: 2006929197

CR Subject Classification (1998): C.2, E.3, F.2, H.4, D.4.6, K.6.5

LNCS Sublibrary: SL 5 – Computer Communication Networks and Telecommunications

ISSN   0302-9743
ISBN-10   3-540-36646-6 Springer Berlin Heidelberg New York
ISBN-13   978-3-540-36646-1 Springer Berlin Heidelberg New York

Springer is a part of Springer Science+Business Media

springer.com

© Springer-Verlag Berlin Heidelberg 2006
Printed in Germany

Typesetting: Camera-ready by author, data conversion by Scientific Publishing Services, Chennai, India
Printed on acid-free paper   SPIN: 11801412   06/3142   5 4 3 2 1 0

Preface

The 2005 Secure Mobile Ad-hoc Networks and Sensors (Secure MADNES 2005) International Workshop was the first of an annual series of conferences on security issues for emerging technologies. Its purpose is to create a collaborative research forum bringing together the mobile networking community and the security community.

Secure MADNES 2005 received 33 submissions. The submitted papers were carefully reviewed, each by at least three members of the Program Committee. The Committee selected 12 papers for inclusion in the conference preproceedings, resulting in an acceptance rate of 36%. Revised versions of the accepted papers are included in these proceedings.

In addition to the presentations of the submitted papers, the conference program included five invited keynote lectures by eminent researchers, and a panel discussion. The Datamaxx Group Keynote was given by Virgil Gligor, on the security of emergent properties in traditional ad-hoc networks. Rebecca Wright gave the second keynote lecture on privacy-preserving data-mining. The third keynote lecture was by Evangelos Kranakis on enhancing intrusion detection in future wireless mobile networks. This was followed by a keynote lecture by Yuliang Zheng on efficient cryptographic techniques for mobile ad-hoc networks. The last speaker, René Peralta, could not be present at the meeting (having just joined NIST, the National Institute for Standards and Technology). His invited paper, on determining network topology properties via dark-encounter computations, is included in the proceedings.

The Panel Discussion focused on authentication in constrained environments, and featured the participation of Virgil Gligor, Evangelos Kranakis, Doug Tygar, Yuliang Zheng, and Mike Burmester, in addition to enthusiastic participation by conference attendees.

We thank the Program Committee (listed on the next page) for their hard work in reviewing the papers and selecting the program from the submitted papers. We also thank Feng Bao and Breno de Medeiros for their work as General Chairs, Breno de Medeiros for compiling the transcripts of our panel, and Tri van Le for checking and compiling these proceedings.

We are delighted to thank the generous financial support from the Army Research Office, the Datamaxx Group, and the FSU SAIT Laboratory.

Finally, we thank all the authors who submitted papers, all the conference attendees, our keynote speakers and our panelists, whose participation made the conference possible.

May 2006

Mike Burmester
Alec Yasinsac

MADNES 2005
September 20-22, 2005, Singapore

MADNES 2005 was organized by the SAIT Laboratory and the U.S. Navy Research Office in cooperation with ISC 2005 and IWAP 2005.

General Chairs
Feng Bao
Institute for Infocomm Research, Singapore
Breno de Medeiros
Florida State University, Florida, USA

Program Chairs
Mike Burmester
Florida State University, Florida, USA
Alec Yasinsac
Florida State University, Florida, USA

Program Committee

N. Asokan .... Nokia Research Center, Finland
Giuseppe Ateniese .... Johns Hopkins University, USA
Feng Bao .... i2r, Singapore
John Baras .... University of Maryland, USA
Sonja Buchegger .... EPFL LCA, Switzerland
Levente Buttyan .... Budapest University of Tech. and Eco., Hungary
Stephen Carter .... Raytheon Corporation, USA
Claude Castelluccia .... INRIA, France
Bruce Christianson .... University of Hertfordshire, UK
Jim Davis .... Iowa State University, USA
Breno de Medeiros .... Florida State University, USA
David Evans .... University of Virginia, USA
Virgil Gligor .... University of Maryland, USA
Zygmunt Haas .... Cornell University, USA
Fritz Hohl .... Sony WSL, Germany
Markus Jakobsson .... University of Indiana Bloomington, USA
Kwanjo Kim .... Information and Comm. University, Korea
Panayiotis Kotzanikolaou .... University of Piraeus, Greece
Evangelos Kranakis .... Carleton University, Canada
Tri Van Le .... Florida State University, USA
Javier Lopez .... University of Malaga, Spain
Chris Mitchell .... Royal Holloway University of London, UK

Rene Peralta .... Yale University, USA
Avi Rubin .... Johns Hopkins University, USA
Christian Tschudin .... UNIBAS, Switzerland
Tsutomu Matsumoto .... Yokohama National University, Japan
Giovanni Vigna .... University of California at Santa Barbara, USA
Cliff Wang .... Army Research Office, USA
Andre Weimerskirch .... University of Bochum, Germany
Susanne Wetzel .... Stevens Institute of Technology, USA
Moti Yung .... Columbia University, USA
Yuliang Zheng .... University of North Carolina, Charlotte, USA

Sponsors

The U.S. Army Research Office
North Carolina, USA.

The SAIT Laboratory
Florida State University, USA.

The Datamaxx Group
Tallahassee, Florida, USA.

Table of Contents

Mobile Ad-Hoc Networks and Sensors

Keynote: On the Security of Emergent Properties in Traditional and Ad-Hoc Networks

Virgil D. Gligor

Electrical and Computer Engineering Department
University of Maryland, College Park, Maryland 20742
`gligor@beckmann.eng.umd.edu`

Abstract. A common characteristic of all networks is that of emergent properties. Intuitively, emergent properties are features that cannot be provided by individual network nodes themselves but instead result from interaction and collaboration among network nodes. In this paper, we present the salient characteristics of these properties (e.g., they are probabilistic, their locus and time may be uncertain) and discuss their security implications. Several examples of emergent properties in sensor and ad-hoc networks are discussed including key connectivity, trust establishment, and node replica detection. We contrast these examples with those of more traditional networks (e.g., distributed denial of service, and point out the inherent limitations of the end-to-end argument which guided much of the Internet design, in network security. We conclude with a common theme of current research in security of emergent properties, namely that of a new threat model whereby the adversary may adaptively compromise nodes of a network.

M. Burmester and A. Yasinsac (Eds.): MADNES 2005, LNCS 4074, p. 1, 2006.

A Novel Pairwise Key Predistribution Scheme for Ubiquitous Sensor Network

Jong Sou Park[1], Mohammed Golam Sadi[1], Dong Seong Kim[1],
and Young Deog Song[2]

[1] Computer Engineering Department, Hankuk Aviation University, Korea
{jspark, sadi, dskim}@hau.ac.kr
[2] Department of Information Security Engineering, Hanseo University, Korea
songyd0614se@naver.com

Abstract. Secure communication is an open challenge for Ubiquitous Sensor Network (USN) collecting valuable information. Sensor nodes have highly constrained resources like limited battery power, memory, processing capabilities etc. These Limitations make infeasible to apply traditional key management techniques such as public key cryptography or other complex cryptographic techniques in the USN. Based on polynomial key pre-distribution protocol, we propose a novel key predistribution technique to establish a pairwise key between sensors. In our scheme, we adopt the Probabilistic Randomness concept on the basis of Grid Based Scheme in a way to achieve improved resiliency over large number of node compromise compared to the existing key predistribution schemes. Beside substantial improved resiliency, it also provides high probability and efficiency in establishing pairwise keys both in direct or path discovery method. Security analysis shows the effectiveness of our scheme in terms of resiliency improvement with little additional overheads in memory and communication.

1 Introduction

In this paper, we propose a novel key predistribution scheme for Ubiquitous Sensor Networks (USN). USN consists of a large collection of small autonomous devices called sensor nodes. A sensor node is powered with battery and equipped with integrated sensors. It has limited data processing capabilities and can perform short-range radio communication with other sensors. Resource constraints in the sensor node establish the perception that traditional Public Key Cryptography (PKC) and key distribution center are beyond the capabilities of sensor nodes. The software implementation of ECC for 8-bit CPUs has showed the possibility to take advantage of PKC to constrained devices such as embedded system [6], but it is still not feasible for USN since sensor nodes are highly resource constrained than embedded system. In order to ensure security to USN, we exploit key predistribution schemes that have been widely accepted since the early stage of Wireless Sensor Network (WSN) development. USN uses the infrastructure of WSN and consists of a large number of sensor noses. Accordingly, the key predistribution schemes can be applied to USN. Several studies

M. Burmester and A. Yasinsac (Eds.): MADNES 2005, LNCS 4074, pp. 2–13, 2006.
© Springer-Verlag Berlin Heidelberg 2006

[1 − 4, 6, 7] have proposed key predistribution schemes for securing WSN but they have some drawbacks against large number of node comprise which has the same effect on USN. In order to solve the problem of resiliency against large number of node compromise, we propose a novel key predistribution scheme. In our scheme, we adopt probabilistic randomness concept on the basis of the grid based scheme. The combined effect of these two basic schemes improves the network resiliency to a much higher level against large number of node compromise. Besides this improvement, our scheme also ensures establishment of pairwise keys between any pair of nodes with a high probability. Even if some nodes in the network are compromised, the scheme ensures the establishment of pairwise keys between the non-compromised nodes. Another advantage is that a sensor can determine whether it can establish a pairwise key with another node, which reduces communication overhead in some extent.

2   Overview of Our Novel Pairwise Scheme

This section presents our proposed scheme in detail. We use grid based scheme [7] as the basis of our scheme and then imply the concept of the probabilistic randomness [4] over the grid based scheme. For a sensor network of N nodes, the setup server constructs a $m \times m$ grid where $m = \lceil \sqrt{N} \rceil$ and allocates distinct group of polynomials to each row and column of the grid. The setup server then assigns each sensor node an intersection in the grid. For the sensor at the intersection (i, j), τ polynomials are randomly chosen from each polynomial groups assigned to the corresponding row and column. If any two nodes have polynomial shares in common, they can establish a unique pairwise key between them. We can choose the the value of N larger than the actual number of sensor nodes to be able to expand the network in future. We denote the ID of a sensor as the concatenation of the binary representations of the row and column coordinates $[l = \lceil log_2 m \rceil]$. Syntactically, we represent an ID constructed from the coordinate (i, j) as $\langle i, j \rangle$. For ease of presentation, we denote ID i as $\langle r_i, c_i \rangle$, where r_i and c_i is the first and last l bits of i respectively.

Pairwise key establishment between two sensor nodes is performed in a sequence of three phases: Setup, direct key establishment and path key establishment. In the setup phase polynomial shares are distributed to the sensor nodes before deployment. After deployment, sensors start second phase to establish a direct pairwise between neighbors. If they can establish a pairwise key successfully then they do not need to go through the third phase. Otherwise, they need to start path key establishment phase to establish a path key by the help of other sensors in a path. The details of each phases are explained below.

2.1   Setup: Polynomial Assignment

The setup server does the following activities to initialize the sensors for the network:

- Randomly generates $2m\omega$ number of t degree bi-variate polynomials $F = \{f_i^r(x, y), f_i^c(x, y)\}_{i=0,...,m\omega-1}$ over a finite field F_q, where q is prime number that is large enough to accommodate a cryptographic key.

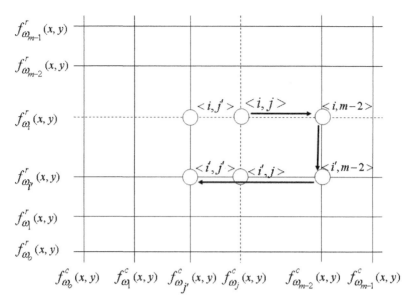

Fig. 1. Polynomial allocation and key discovery mechanism of our scheme

- To identify the polynomials the setup server assigns each polynomial a unique ID.
- Groups the polynomials $f_i^r(x,y)_{i=0,...,m\omega-1}$ into m subgroups $f_{\omega_i}^r$ $(x,y)_{i=0,...,m-1}$, where each group contains ω distinct polynomials and assigns to each row of the grid.
- Similarly the setup server groups the polynomials $f_i^c(x,y)_{i=0,...,m\omega-1}$ into m distinct subgroups $f_{\omega_i}^c(x,y)_{i=0,...,m-1}$ that are allocated to each column of the grid as shown in Figure 1.
- The setup server allocates an unoccupied intersection (i,j) in the grid for each sensor and randomly picks τ polynomials from the two allocated polynomial groups corresponding to i^{th} row and j^{th} column of that intersection. Then these polynomial shares of these 2τ polynomials with their IDs and the ID $\langle i,j \rangle$ are assigned to the sensor node.
- The possible order of allocation of sensor nodes in the grid is done in a way shown in Fig. 2 so that they will occupy the dense rectangular area to facilitate the path discovery phase.

2.2   Direct Key Establishment

Direct key establishment between any two sensor nodes is accomplished in two steps. Firstly, the sensors need to find out whether they are able to establish a pairwise key between them. Secondly, they need to find the exact common polynomial share of the same bivariate polynomial to establish a pairwise key. These two steps are explained below:

- The sensors broadcast their IDs in simple text. After getting an ID from its neighbor it checks the row and column part of the ID. For example, any two

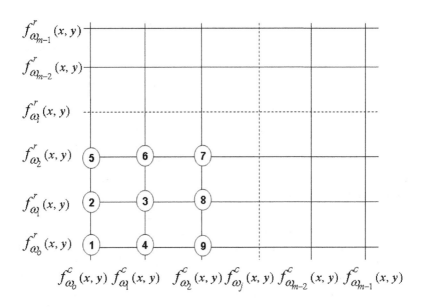

Fig. 2. An example of order of node assignment in our scheme

sensor nodes i and j check whether $c_i = c_j$ or $r_i = r_j$. If $c_i = c_j$, then it is confirmed that both nodes have τ polynomial shares from the polynomial group $f_{\omega_i}^c(x, y)$. Similarly if $r_i = r_j$, then their polynomial share are selected from the same polynomial group $f_{\omega_i}^r(x, y)$.

- In second step, to discover a common bivariate polynomial, these two sensors exchange the IDs of their τ polynomial shares by broadcasting them in simple text. To protect the IDs from the attacker, one node may challenge the other node to solve puzzles instead of disclosing the IDs of the polynomials directly. For example, using the challenge response protocol [2, 4], sensor node i may broadcast an encrypted list α, $E_{k_v}(\alpha)_{v=1,2...,\tau}$, where k_v is a potential pairwise key the other node may have by evaluating the polynomial shares. If node j can correctly decrypt any one of these, it can establish pairwise key with the source node i.

2.3 Path Key Establishment

If two sensors fail to establish a pairwise key directly, they must start path key establishment phase. For example, nodes i and j have to use path discovery when $c_i \neq c_j$ and $r_i \neq r_j$ or when they do not have any common polynomial share on the same bi-variate polynomial. We note that either node $\langle r_i, c_j \rangle$ or $\langle r_j, c_i \rangle$ has a high probability to establish a pairwise key with both the nodes i and j as they have τ polynomials selected randomly from same polynomial group. Indeed, if there is no compromised node, it is obvious that there might exist at least one node that can be used as an intermediate node between any two sensor nodes due to the node assignment algorithm. For example, in Fig. 1, both nodes $\langle i',j \rangle$ and $\langle i,j' \rangle$ can help node $\langle i,j \rangle$ to establish a pairwise key with node $\langle i',j' \rangle$. However, in some situations, if both of the above intermediate nodes have been compromised or out of communication range, there are still alternative key paths. For example, in Fig. 1, nodes $\langle i,m-2 \rangle$ and $\langle i',m-2 \rangle$ can work together to help node $\langle i,j \rangle$ to setup a common key with $\langle i',j' \rangle$. Thus all the nodes that belong to same row or column of the of the communicating nodes, can help to setup a pairwise key between them. Indeed, there are up to $2(m-2)$ pairs of such nodes in the grid. The algorithm to discover key paths between sensor nodes S and D using two intermediate nodes is same as proposed in [7] and stated as follows:

- The source node S determines a set L of non-compromised nodes with which it can establish pairwise key directly using a non-compromised polynomial. S randomly picks a set L_d of d sensor nodes from L. S also generates a random number r, and maintains a counter c with initial value 0.
- For each node $u \in L_d$, S increments the counter c and computes $K_c = F(r,c)$, where F is a pseudo random function [5]. Then S sends to u the IDs of S and D, c and K_c in a message encrypted and authenticated with the pairwise key $K_{s,u}$ between S and u.
- If a sensor node $u \in L_d$ receives and authenticates such a message, it knows that node S wants to establish a pairwise key with D. Node u then checks whether the two sensor nodes $\langle c_u, r_d \rangle$ and $\langle c_d, r_u \rangle$ are compromised or not. If u finds a non-compromised node v, u can establish a pairwise key with D through v. Then u sends the IDs of S and D, c and K_c to v in a message encrypted and authenticated with the pairwise key $K_{u,v}$ between u and v.

3   Security Analysis of Our Scheme

In our scheme, each sensor has 2τ polynomial shares where each τ polynomial shares are randomly taken from the distinct subgroups allocated to the corresponding row and column. We assume there are $N = m \times m$ sensors in the network. According the node assignment algorithm, each sensor node has a very high probability to establish a pairwise key with $2(m-1)$ sensor nodes directly. Thus, among all the other sensors, the percentage of nodes that a node can establish a pairwise key directly is,

$$\frac{2(m-1)}{N-1} \approx \frac{2(m-1)}{m^2-1} = \frac{2}{m+1} \tag{1}$$

First we compute the desired values for security parameter and then focus our attention to the performance of the our scheme under two types of attacks. First, the attacker may target the pairwise key between any two sensors. The attacker may either try to compromise the pairwise key or prevent the two sensor nodes from establishing a pairwise key. Second, the attacker may target the entire network to lower the probability that two sensors may establish a pairwise key or to increase the cost to establish pairwise keys.

3.1 Preferred Values of ω, τ

For any pair of nodes, establishing a pairwise key between them is possible if the key sharing graph of the nodes is connected. Given the size and the density of a network, we can calculate the values for ω and τ so that the node graph is connected with high probability. We use the following approach adapted from [4].

Computing Required Local Connectivity. Let p_c be the probability that the key sharing graph is connected. We call it global connectivity. We use local connectivity to refer to the probability of two neighboring nodes sharing at least one polynomial. To achieve a desired global connectivity p_c, the local connectivity must be higher than a certain value; we call this value the required local connectivity, denoted by $p_{required}$. Using connectivity theory in a random graph by Erdos and Renyi [8], we can obtain the necessary expected node degree d (i.e. the average number of edges connected to each node) for a network of size N, when N is large in order to achieve a given global connectivity, p_c:

$$d = \frac{N-1}{N}\left[ln\left(N\right) - ln\left(-ln\left(p_c\right)\right)\right] \tag{2}$$

For a given density of sensor network deployment, let n be the expected number of neighbors within wireless communication range of a node. Since the expected node degree must be at least d as calculated above, the required local connectivity $p_{required}$ can be estimated as:

$$p_{required} = \frac{d}{n} \tag{3}$$

Computing Actual Local Connectivity. After we have selected values for ω and τ, the actual local connectivity is determined by these values. We use p_{actual} to represent the actual local connectivity, namely p_{actual} is the actual probability of any two neighboring nodes sharing at least one polynomial share (i.e. they can find a common key between them). Since $p_{actual} = 1 - p_r$ (two nodes do not share any polynomial),

$$p_{actual} = 1 - \frac{\binom{\omega}{\tau}\binom{\omega-\tau}{\tau}}{\binom{\omega}{\tau}^2} = 1 - \frac{((\omega-\tau)!)^2}{(\omega-2\tau)!\omega!} \tag{4}$$

The values p_{actual} have been plotted in Fig. 3 when ω varies from 0 to 100 for values of $\tau = 2, 4, 6, 8$. For example, when $\tau = 6$, the largest ω we can choose while achieving the local connectivity $p_{actual} \geq 0.7$ is 35.

Computing ω and τ. Getting the required local connectivity $p_{required}$ and the actual local connectivity p_{actual}, in order to achieve the desired global connectivity p_c, we should have $p_{actaul} \geq p_{required}$,

$$1 - \frac{((\omega-\tau)!)^2}{(\omega-2\tau)!\omega!} \geq \frac{N-1}{nN}\left[ln(N) - ln(p_c)\right] \tag{5}$$

Therefore, in order to achieve certain p_c for a network of size N and the expected number of neighbors for each node being n, we just need to find values of ω and τ, such that Inequality (5) is satisfied.

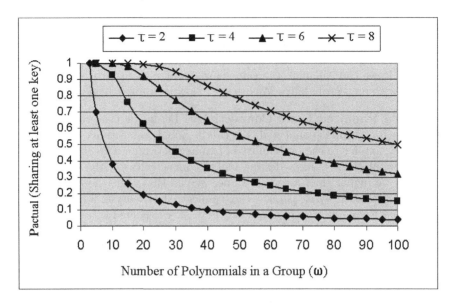

Fig. 3. Probability of sharing at least one key between nodes

3.2 Attacks Against a Pair of Sensors

One simple way to compromise a pairwise key is to compromise one of the related sensors. In this case, security protocol has nothing to do. The other way to compromise the pairwise key without compromising the related nodes is to compromise the shared polynomial between them. Our scheme provides a strong resistance for this situation. Assume two nodes u and v can establish a pairwise key directly. To compromise the shared pairwise key between them, the attacker needs to compromise at least $t + 1$ sensors that have polynomial shares on same bivariate polynomial. Due to the polynomial distribution algorithm only the sensor nodes that belong to same row or column as the related sensors, has a certain probability to have the same polynomial shares. So, it is very difficult to compromise the selected $t + 1$ sensors from the pool of nodes in the network. Further, we may abolish this threat by restricting the use of each polynomial shares at most $t + 1$ times among the sensors in a particular row or column in the grid. As a result, an attacker cannot recover a polynomial unless he/she compromises all the related sensors. This method puts a limit on the maximum numbers of sensors in a particular row or column for a combination of ω and τ . Indeed given the above constraint, the total number of sensors in a row or column cannot exceed $\frac{(t+1) \times w}{\tau}$. Even if the attacker successfully compromises the polynomial (as well as the pairwise key), the related sensors can still reestablish pairwise key using one of the noncompromised polynomial shares stored in their memory or using new round of path discovery process. From the path discovery process, we know that there are $m - 1$ pair of nodes, which can help u and v to reestablish a pairwise key. To prevent node u from establishing a key with node

v completely attacker has to compromise all of the pair of nodes otherwise there will be a possibility to establish a pairwise key between them through multiple rounds of path discovery process.

Now we consider the scenario where nodes u and v establish a pairwise key through path key establishment. The attacker may compromise one of the sensors involved in the key path. If the attacker has the message used to deliver the key, he/she can recover the pairwise key. However, the related sensors can establish a new key with a new round of path key establishment once the compromise is detected. To prevent the sensors from establishing another pairwise key, the attacker has to block at least one sensor in each path between u and v. There are $2m - 2$ key paths between u and v that involve one or two intermediate nodes. Besides the key path with the compromised node, there are at least $2m - 3$ paths. To prevent pairwise key establishment, the attacker has to compromise at least one sensor in each path.

3.3 Attacks Against the Network

Having the knowledge of the subset assignment mechanism, adversary may compromise the bivariate polynomials in F one after another by compromising selected sensor nodes to compromise the whole network. Suppose the adversary just compromised l bivariate polynomials in F. In the worst case, ml sensors in the grid have a probability that one of their polynomial shares is compromised if each polynomial belongs to different row or column. But those nodes can work by establishing new pairwise key using one of the remaining non-compromised polynomial shares stored in their memory or using new round of path discovery process. In the above case, there are $(m - l)m$ sensors whose polynomial shares are uncovered and key establishment process between them will not be affected. Thus, the attacker compromises about $(t + 1)l$ sensors but these may affect only the pairwise key establishment of ml sensors with a small probability. However we see that adoption of randomness in the Blom's scheme [3] enhances the resiliency to node capture. Now we will calculate the probability of at least one key is disclosed when nodes are randomly captured. It follows from the security analysis in [1] that an attacker cannot determine noncompromised keys if he or she has compromised more than t sensor nodes. We assume that an attacker randomly compromises N_c sensor nodes, where $N_c > t$. Consider any polynomial f in F. The probability of f being chosen for a sensor node is $\frac{\tau}{m\omega}$ and the probability of this polynomial being chosen exactly k times among N_c compromised sensor nodes is,

$$P(k) = \frac{N_c!}{(N_c - k)!k!}\left(\frac{\tau}{m\omega}\right)^k\left(1 - \frac{\tau}{m\omega}\right)^{N_c-k} \tag{6}$$

Thus, the probability of any polynomial being compromised is $P_c = 1 - \sum_{k=0}^{t} P(k)$. Since f is a polynomial in F, the fraction of compromised links between noncompromised sensors can be established as P_c.

Fig. 4 shows the relationship between fraction of compromised links for noncompromised sensors and number of compromised nodes for a combination of

$\omega = 25$ and $\tau = 6$. According to the graph, we see that our scheme has a very high resiliency when a large number of the sensor nodes are compromised. For example, in the case of a sensor network of 20,000 nodes, if the attacker compromises 40% of the total nodes (i.e. 8,000 nodes) then only about 5% of the links of noncompromised nodes are affected. Thus, the majority of the noncompromised nodes are not affected.

3.4   Comparison with Previous Schemes

Let us compare the resiliency of our scheme with basic probabilistic scheme, q-composite scheme and grid based scheme. Here we assume the network size N is 20000, $m = 142$ and the probability $p = 0.24$. In Fig. 4, the four curves show fraction of compromised link as a function of number of compromised sensor nodes. Basic probabilistic scheme has almost same performance as the q-composite scheme ($q = 2$) and the grid based scheme works well up to 2000 compromised nodes. In contrast, our scheme provides sufficient security up to 9000 compromised nodes and then the performance gradually decreases. Here we assume the value of the security parameters $\omega = 25$, $\tau = 6$ and degree of polynomial $t = 19$.

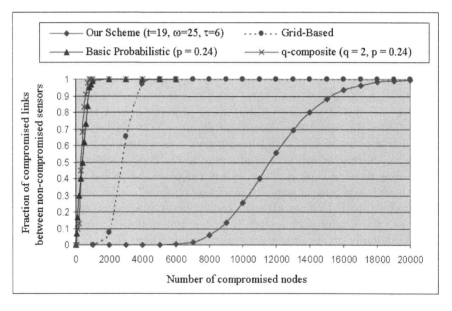

Fig. 4. Fraction of compromised links between noncompromised sensors vs. number of compromised sensor nodes

3.5   Overheads of Our Scheme

This scheme has reasonable memory requirements mainly for storing several polynomials chosen randomly. Each sensor needs to store 2τ polynomials of degree t with their IDs. Assume b bits are required to represent a polynomial ID. Then we can write $b = log_2 (2m\omega)$. In addition, a sensor needs to store the IDs of the compromised nodes with which it can establish a pairwise key directly. Thus, the total storage overhead in each sensor is at most

$$Memory = 2\tau (t + 1) logq + 2\tau b + 2 (t + 1) l \qquad (7)$$

According to grid based scheme [7], the storage overhead in each sensor is at most $(t + 1) logq + 2 (t + 1) l$. A comparison of these two equations shows that our scheme requires almost τ times more memory than the grid-based scheme. Memory overhead of our scheme mainly depends on the of the security parameter τ and polynomial degree t. Fig. 3 describes how the probability of sharing at least one key among nodes varies for different values of and τ. But we can regulate the values of τ, ω and t to adjust the network security with sufficient memory accommodation. To get the desired security shown in Fig. 4 each sensor needs the storage capacity, which is equivalent to store almost 260 keys as used in basic probabilistic scheme. Though it needs more memory, it is a not significant factor for sensor network because sensor memory size will be increased in near future as technology is developing very fast.

In terms of communication overhead, our scheme has additional overhead compared to the grid based scheme due to broadcasting of polynomial IDs during direct key establishment process. But this additional communication process will occur only once during the initialize period of the network. Also the path discovery process that uses two or more intermediate nodes introduces additional communication overhead similar to grid based scheme due to exchange of several handshaking signals among the sensors in the path. The computational overhead is essentially the evaluation of one or multiple t-degree polynomials that can be done efficiently using the same approach as in [7].

4   Related Works

The key predistribution schemes have been proposed in several literatues [2, 4, 7] for WSN. For the first time, Eschenauer and Gligor have proposed the basic probabilistic scheme [4]. The main idea is to let each sensor node randomly pick a set of keys from a key pool before deployment. Any two nodes have a certain probability to share at least one key that act as the secret key between them. Chan et. al. further extended this idea and developed two key predistribution schemes [2]: q-composite key scheme and random pairwise keys scheme. The q-composite key scheme also uses a key pool but requires two sensors compute a pairwise key from at least their shared q-predistributed keys. The random pairwise keys scheme picks pair of sensors and assigns each pair a unique random key. Both these schemes improve the resiliency over basic probabilistic scheme. However the basic probabilistic and q-composite key scheme provides very poor performance when the number of com-promised nodes increases. The random pairwise keys scheme overcomes the above problem but it needs much memory requirement. Liu and Ning have proposed two efficient schemes [7]: random subset assignment and grid based key predistribution scheme that have basis of polynomial key predistribution. In grid based scheme a conceptual grid is formed and a unique polynomial function is allocated to each row and column of the grid. A sensor node is allocated to a particular intersection of the grid so the two polynomials corresponding to that row and column are assigned to its memory. If any two sensors have same column or row number then certainly they can establish a pairwise key. This scheme has a number of nice facilities such as high probability to establish pairwise keys, resiliency to node capture, low communication overhead and reduced computation in the sensor node. However, the resiliency to node capture is not acceptable when the compromise of nodes grows larger than certain threshold value.

Our scheme has not only robust resiliency against large number of node compromise compared to previous approaches but also flexibility to adjust security level according to the desired requirement by regulating the security parameters. We can explicitly say that our scheme has a substantial improvement in network resiliency with little memory overhead and reasonable communication and computation workload. Besides, our scheme provides a high probability to establish of pairwise key between two nodes directly or via intermediate nodes and also provides efficiency in determining the path key. Even if some nodes are compromised, there is still a high probability to establish a pairwise key between non-compromised nodes. Finally, this scheme allows optimized deployment of sensors to establish pairwise key directly due to orderly assignment of grid intersections.

5   Concluding Remarks

We have presented a novel pairwise key distribution scheme for USN. The analysis of our scheme depicts the feasibility to apply this scheme to USN compared to the existing schemes. It provides significant improvement in resiliency compare to the existing key management schemes over a large number of node compromises. The probability of establishing a pairwise key between nodes is very high and our scheme enables to adjust the security level by regulating the security parameters. Pairwise key establishment using direct method or path key discovery method is efficient with a little memory and communication overheads compare to other existing schemes but these overhead is endurable in sensor nodes. In future works, we will perform network wide security analysis of our proposed scheme through simulation method and detailed analytic methods.

Acknowledgement

This research was supported by the Internet information Retrieval Research Center (IRC) in Hankuk Aviation University. IRC is a Regional Research Center of Gyeonggi Province, designated by ITEP and Ministry of Commerce, Industry and Energy.

References

1. Blando, C., Saints, De, A., Herzberg, A., Kutten, S., Vaccaro, U., Yung, M.: Perfectly-Secure Key Distribution for Dynamic Conferences. Lecture Notes in Computer Science, Vol. 740. Springer-Verlag, Berlin Heidelberg New York (1993) 471-486

2. Chan, H., Perrig, A., Song, D.X.: Random Key Predistribution Schemes for Sensor Networks. In Proc. of the 2003 IEEE Sym. on Security and Privacy (2003) 197

3. Du, W., Deng, Jin., Han, S.Y., Varshney, P.K.: A Pairwise Key Pre-distribution Scheme for Wireless Sensor Networks. In Proc. of the 10th ACM conf. on Computer and Communications Security. (2003) 42-51

4. Eschenauer, L., Gligor, V.D.: A Key-Management Scheme for Distributed Sensor Networks. In Proc. of the 9th ACM Conf. on Computer and Communications Security. (2002) 41-47

5. Goldreich, O., Goldwasser, S., Micali, S.: How to Construct Random Functions. J. of the ACM. Vol. 33. Issue 4 (1986) 792-807

6. Gupta, V., Millard, M., Fung, S., Zhu, Yu., Gura, N., Eberle, H., Shantz, S.C.: Sizzle: A Standards-Based End-to-End Security Architecture for the Embedded Internet. In Proc. of 3th IEEE Int. Conf. on Pervasive Computing and Communications. (2005) 247-256

7. Liu, D., Ning, P.,:Establishing Pairwise Keys in Distributed Sensor Networks. In Proc. of the 10th ACM Conf. on Computer and Communications Security. (2003) 52-61

8. Erods, P., Renyi, A.: On Random Graph. Publicationes Mathematicae. (1959) 290-297

Key Management for Mobile Sensor Networks

David Sánchez Sánchez and Heribert Baldus

Philips Research, Connectivity Systems Group
52066 Aachen, Germany
{david.s.sanchez, heribert.baldus}@philips.com

Abstract. Key management is paramount for mobile sensor network (MSN) security. The special nature of MSNs imposes challenging requirements on key management design. We evaluate current key management techniques proposed for wireless sensor networks in the context of MSNs and identify open issues for future research. We also propose a novel approach to control and replace pool-based pre-distributed keys. The analysis in this paper shows that this approach keeps the initial connectivity and resiliency properties of the key pre-distribution scheme even when nodes and keys are revoked.

1 Introduction and Motivation

Wireless sensor networks (WSNs) have received a lot of attention for military and civilian applications [18]. WSN nodes may be left unattended and the WSN may be formed in public or hostile areas where communication is monitored and sensor nodes are subject to capture and manipulation by an adversary.

In this paper, we explore specific security issues of MSNs. In difference to currently considered WSNs, MSNs possess special operational requirements that impose even more challenging constraints on security design than those placed by to date considered WSNs. Security services for MSN include authentication and communication confidentiality and integrity. Key management is fundamental for MSN security.

The computational and communication constraints of MSN sensor nodes make unfeasible to use any solution based on public key cryptography. The mobile ad hoc nature of MSNs makes typical online server-based solutions unpractical. Key management based on key pre-distribution schemes (KPS) is the best option for MSNs. However, unknown MSN membership and size, absence of post-deployment information as well as mobility of nodes enforce very strict requirements on KPSs.

The contributions of this paper are two fold. Firstly, we exhaustively evaluate existing KPS proposals in the context of MSN, identify their limitations and provide directions for future research. Secondly, we present a novel approach to control and replace pool-based pre-distributed keys in MSNs. This system enables keeping the initially required connectivity and resiliency for MSNs, even when sensors erase keys from their key rings in the event of node revocations.

The remainder of this paper is organized as follows. In Section 2, we provide a classification of WSNs. In Section 3, we review and evaluate key management

M. Burmester and A. Yasinsac (Eds.): MADNES 2005, LNCS 4074, pp. 14–26, 2006.

approaches in the context of MSNs. Section 4 presents a system to control and replace pool-based pre-distributed keys and Section 5 evaluates the performance of the proposed key replacement system. Finally, we conclude this paper in Section 6.

2   Classification of Wireless Sensor Networks

We classify WSNs in two types: distributed WSN (DSN), and mobile WSN (MSN). A DSN is assumed to be a large-scale, wireless, ad hoc, multi-hop, unpartitioned network of tiny, mostly *immobile* (after deployment) sensor nodes. The number N of DSN nodes can range up to thousands of nodes or even more. In some applications [18,11,12], while the number N of system nodes may still be huge, DSN-wide connectivity is not required or cannot be guaranteed (e.g. in WSNs composed of very short transmission range mobile sensor nodes). In a system with N nodes, there may be multiple independent WSNs connecting different subsets of the N nodes.

A MSN is defined as a small, medium or large scale WSN of *mobile*[1] (after deployment) sensor nodes. After deployment, mobile nodes may spontaneously join or quit the MSN. A priori the size and membership of a MSN is unknown: the MSN nodes may be present at the moment of the MSN initial formation or may join and quit subsequently. For comparison, a number of MSNs can be seen as multiple disconnected partitions of a large-scale DSN.

In MSN applications, a large group of interoperable mobile nodes (thousands of nodes) are initially deployed in an area and then, by different means, spontaneously join to form different MSNs, i.e. the MSN is typically composed of a subset of the N nodes. New nodes may be deployed in subsequent phases to the initial deployment. Since the MSN is mobile, base stations (BS), which collect sensor data, are also typically mobile. The most practical situation is that mobile BSs (MBS) regularly connect to the MSN at given intervals in time, without the need for MBSs to be continuously available at the MSN.

Sensor nodes are battery-powered, have limited computational capabilities and memory, and rely on intermittent wireless communication. Sensors are attached to mobile objects (e.g. to a human body).

3   Evaluation of Key Management Approaches

Public key is unfeasible on sensor nodes due to memory and computational limitations. Therefore, we do not review any public key management approach. We review and evaluate key management techniques for distributing symmetric keys under the context of MSN.

3.1   Evaluation Criteria

For MSN security design, we know the total number N of system nodes but we cannot foresee the size and membership of a MSN (typically from 2 to tens,

---

[1] Nodes have locomotion capabilities.

hundreds or thousands of nodes). We assume no tamper-resistant sensor nodes. To evaluate key management techniques for MSN we consider the following factors: (1) scalability, (2) robustness (resistance to node captures and other security attacks such as man-in-the-middle), (3) computational and communication performance and storage needs, (4) feasibility for mobile ad hoc networking, (5) connectivity properties in small/medium/large MSNs, and (6) security services enabled (confidentiality, integrity and authentication).

3.2   Server-Based Key Management

Perrig et al. [17] proposed SPINS, a security architecture specifically designed for WSNs. In SPINS, each sensor node shares a secret key with a BS. Two arbitrary sensor nodes need to use the BS as a trusted third party (TTP) to set up a common secret key.

Key establishment through a TTP can be applied to MSNs only under the following assumptions: (a) there exist a MBS per MSN, (b) MBS is always available at the MSN, (c) each and every MBS shares a secret key with the N system nodes, and (d) the MSN is of small size. However, this solution has very important drawbacks for MSN security. Firstly, each MBS becomes a single point of compromise and failure. Secondly, in dynamic scenarios, where MSNs split and move apart, the number of MBSs must dynamically increase. Thirdly, key management is not scalable in medium/large-scale MSNs.

3.3   System Key Pre-distribution

A KPS enables nodes of a WSN to establish keys independently of any base station. The simplest KPS approach is to pre-load a single system key onto all nodes before deployment. This approach can be applied to MSN security and is optimal on memory cost on nodes. However it does not enable node unique authentication and has no resiliency against node captures unless tamper-resistance is assumed. Zhu et. al. [8] propose to use a system key to establish pairwise keys at the initial phase of sensor deployment in DSNs. This approach cannot be applied to MSNs because it assumes static nodes and DSN-wide connectivity. Additionally, because nodes erase the system key after the initial phase, it does not enable addition of new nodes. Basagni et al. [10] proposed to periodically update the system key. This work assumes tamper-resistant sensor nodes.

3.4   Trivial Key Pre-distribution

In the trivial approach, every sensor shares a pairwise key with each other node of the system. This approach enables direct key establishment and node authentication in MSNs, independently of their membership or size. However, the number of system nodes N allowed is limited by the memory size m available on sensors. In MSN settings, adding new nodes may be difficult.

3.5   Random Key Pre-distribution

Eschenauer and Gligor [3] originally proposed random key pre-distribution for DSNs. The main idea is to load, before deployment, in each sensor u a subset of

m keys (called a *key ring* and denoted K_u) randomly picked from a *key pool S* with $|S|$ keys. In order to establish a pairwise key, two sensor nodes only need to identify the common keys that they (*may*) share. Chan et al. [4] increased the resiliency of Eschenauer's KPS by allowing two sensors to setup a pairwise key only when they share at least n common keys. Di Pietro et al. [9] showed that applying a geometric random model for key pre-distribution further enhances the performance of previous KPSs. Hwang and Kim [7] improved performance of previous KPSs by trading-off a very small number of isolated nodes.

One drawback of the first random KPSs is that, since key rings contain keys randomly picked from the same key pool, more than a pair of nodes may use the same common key to secure their communications. Therefore, these keys cannot be used to provide unique authentication [16].

Following, a number of random KPS proposals, by Chan et al. [4], Du et al. [5] and Liu and Ning [6] enable the establishment of keys that enable unique authentication. Chan et al. pre-distributed random pairwise keys. Du et al. combine Blom's scheme [1] with random key pre-distribution. Liu and Ning [6] applied a similar idea with Blundo's scheme. Similarly, Lee and Stinson [14] combine (m, r, λ, μ)-strongly regular graphs G with a modification on Blom scheme.

The main drawback of random KPSs is that two arbitrary nodes *directly* find a common key with a given probability p (p influences the performance of the KPS). Thus, TTP-based key establishment is often needed. The probability of finding a TTP among the neighbors directly depends on the size of the neighborhood (this, again, influences the performance of the KPS and the network [7]). In small scale MSNs, finding a TTP among the neighbors, may be in many cases impossible.

Additionally, in random KPSs, since each node u is *allowed* (and often required) to work as a TTP, it becomes a single point of compromise for its trusted nodes (similarly to server-based key management). Thus, very advanced efficient intrusion detection systems (IDS), providing reliable node trust and reputation metrics, are required. Alternatively, a key may be established using TTPs in different trust paths, which in turn introduces further communication and computational overhead.

3.6   Deterministic Key Pre-distribution

The risks of using intermediary nodes as TTP is avoided with deterministic KPSs that enable *direct* key establishment to any arbitrary pair of nodes. To this classification belong the system key and trivial key KPSs discussed in sections 3.3 and 3.4.

Blundo's two party [2] polynomial-based KPS provides an interesting system for MSN security. In this KPS, a polynomial share $f(u, y)$, of a symmetric bivariate λ-degree polynomial $f(x, y)$ with coefficients over F_q, is pre-distributed to each sensor u. After the deployment phase, *any* two arbitrary nodes u and v can compute a unique pairwise key $K_{u,v} = f(v, u) = f(u, v)$ by evaluating respective polynomial shares $f(u, y)$ and $f(v, y)$ at the partner node ID.

Blundo's KPS possesses very interesting connectivity properties for MSN, incurs no communication overhead and enables node authentication. It exhibits perfect resiliency up to $\lambda + 1$ captured sensors [2]. Each sensor node u needs memory space m to store $f(u, y)$, which occupies $\lambda + 1$ times the length of a pairwise key of $\lfloor \log q \rfloor$ bits. Blundo's pairwise key establishment can be computationally expensive in sensors with limited CPU capabilities, since it requires λ modular multiplications and λ modular additions in F_q. Liu et al. [6] solve this problem for sensors with low-bit CPUs without division instruction. Each sensor carries t distinct polynomial shares, with coefficients over $F_{q'}$, $q' \ll q$, $t = \lfloor \log q \rfloor / \lfloor \log q' \rfloor$, $q' = 2^k + 1$ (Note q and q' are prime numbers). A pairwise key of $\lfloor \log q \rfloor$ bits is composed by concatenating t partial keys of $\lfloor \log q' \rfloor$ generated from the t polynomial shares. However, this solution decreases the scalability of the original KPS substantially.

Çamtepe and Yener [13] applied combinatorial designs to key pre-distribution. They first propose a simple KPS based on Finite Projective Planes (FPP) with nice properties for MSN security: direct key establishment, tolerance to node captures and no computational or communication overhead. However, it is limited in network scalability and resiliency and it does not enable node authentication. Their hybrid approach augments scalability of the initial scheme to the cost of sacrificing direct connectivity.

Sánchez and Baldus [15] apply an FPP design to the pre-distribution of Blundo polynomial shares. Their approach enables direct pairwise key establishment (and, thus, authentication) for a large number of nodes, independently of the physical connectivity properties of the WSN. Sanchez's KPS exhibits advanced scalability and extremely low computational and communication overhead. The KPS tolerates node captures. However, the resiliency of the scheme against a smart attacker (who captures nodes selectively) is limited by m.

4 Pool-Based Pre-distributed Key Control and Update

One factor that may hinder the application of random/deterministic pool-based KPSs to MSNs is their decreasing connectivity when nodes are revoked. In this section, we propose a system to adapt and improve existing revocation schemes to MSNs.

4.1 Assumptions

Before deployment, keying material is pre-distributed into N sensor nodes using one of the deterministic or random pool-based KPSs discussed in Sections 3.5, and 3.6 (including [3,4,7,9,13]. Initially, each sensor node u carries a key ring K_u with m keys. After deployment, mobile sensors self-organize to form a dynamically changing number of non-interconnected MSNs. Sensors may move from one MSN to another.

Since sensors have limited memory space, sensor data needs to be collected every interval of time T^2. For that reason, we assume the existence of M MBSs,

---

[2] T is typically in the order of hours, days or weeks. In strict real-time applications though, T may be in the order of milliseconds.

which do not necessarily reside at the MSN, where $M \ll N$. In each interval of time T, at least one MBS connects sporadically and shortly to the MSN. Because MBSs are expensive nodes, they are provided of tamper-proof hardware and they are not limited in computing power or memory. Because MBSs are not left unattended and they are typically only present at the MSNs deployment area sporadically, they are not prone to be captured or compromised. We also assume that MBSs have a larger wireless communication range than sensors. Additionally, MBSs can use public key cryptography to secure MBS-to-MBS communications.

4.2   The Problem

Due to the fact that sensors dynamically join different *non-interconnected* MSNs, MSNs are far more vulnerable to attacks from compromised nodes than a full-connected DSN, particularly to node replication or Sybil attacks [19].

A number of intrusion detection systems (IDS) and key revocation approaches have been described for wireless ad hoc networks [3,4,19,20,21,22,24]. The existing key revocation approaches for WSNs use a centralized system [3] or a fully distributed system [4]. In both approaches, once a compromised node c is detected, revocation information is broadcasted to the rest of WSN nodes [3,4]. Key revocation information includes *public votes* associated to c's identifier (ID) [4], a signed list containing the m identities of the keys in c's key ring K_c [3] or *reputation values* on c's identity [24]. In [4] a number t of public votes are needed to definitely revoke a node. After obtaining the sufficient revocation information, each node u firstly verifies if it has shared keys with c, and, secondly, erases the corresponding keys (if any) from its key ring K_u. Note that sensors, which have erased a key in their key pools, have now k, $k < m$, keys in their respective key rings.

These key revocation approaches have two important limitations in the context of MSN. Firstly, a compromised node might be detected and revoked in a MSN but, since MSNs are not interconnected, revocation information cannot be relayed to sensors in other MSNs. How to efficiently and timely distribute this information to the rest of nodes in other MSNs?

Secondly, since the same specific key is carried by many sensors, the fact that revoked keys are erased from sensors decreases KPS connectivity and thus the capacity of nodes to establish secure connections in moving to a new MSN neighborhood, in a newly formed MSNs or in joining other MSNs. For instance, in random pool-based KPSs the probability p that any two nodes share at least one key is a function of the current number k of keys available in each sensor and the total number $|S|$ of keys in the key pool [3]. As k decreases (because keys are revoked), the probability p also decreases (see also Analysis Section). Consequently, the chance of finding sensors with shared keys among a sensor's neighborhood decreases and, in turn, MSN security connectivity decreases.

Sensors dynamically join/quit the MSN or gather together to form new MSNs. If the number k of keys stored in a sensor's key ring decreases progressively nodes cannot secure the network unless each node's neighborhood becomes progressively denser (e.g. by letting sensors move closer). This adaptation, in turn, reduces MSN coverage area and negatively affects network capacity and energy consumption [7]. Moreover, it augments the chance of using undetected adversary-manipulated nodes as TTP for establishing keys.

4.3 Key Revocation and Replacement

To solve the revocation information distribution problem, we propose to exploit the mobility of sensors and MBSs to disseminate revocation information to different MSNs. In some cases, connectivity properties of MBSs can additionally be exploited. To solve the KPS connectivity problem, we propose to replace revoked keys with new keys and to exploit mobility and connectivity properties of MBSs to disseminate new keys to other nodes.

We assume nodes and MBSs have a mechanism to easily and efficiently identify the keys associated to a node. For instance, in some random pool-based KPSs the ID of a sensor and the index of a key in its key ring are used to determine which keys are chosen from the key pool [19]. This method uses extremely computational efficient pseudorandom functions. Thus, it is simple to verify that a key is included in the key ring of a node by using the sensor ID and the index of the revoked key in the node's key ring.

Our system comprises the following phases:

Detection of Node Compromise Locally in a MSN. By using one of the existing IDSs [19,20,21,22,24] a compromised node c is detected at the MSN.

Mobile sensors can disseminate public votes and node trust and reputation information when moving to a different neighborhood of the MSN or when joining another MSN.

Local Key Revocation in a MSN. Each of the sensors, at the MSN where node c is first detected, erases from its key ring the keys matching any of the m keys included in the key ring K_c.

At the moment when a MBS connects to the MSN, local[3] node revocation information is relayed to the MBS. For instance, t public votes or reputation information associated to node c can be relayed by any (or a group) of the non-compromised sensors.

Global Key Revocation and Key Replacement. We assume that a MBS can connect to the rest of MBSs, when possible by using their long-range wireless capabilities or, otherwise, by coming in the vicinity of other MBSs.

Each MBS exchanges its locally-gathered revocation information with the rest of $M - 1$ MBSs. In this manner, all MBS have *global* revocation information, which merges all the information gathered locally and independently at the different MSNs.

After global revocation information is available to all MBSs, they cooperatively identify the keys to be revoked and agree on new keys to replace them. Assume that in the initial key pool with $|S|$ keys, the keys k_i are sequentially

---

[3] The revocation information is gathered locally at the MSN.

identified from 1 to $|S|$. Each m revoked keys $k_{\beta_1}, k_{\beta_2}, \ldots k_{\beta_m}$ are identified by their position in the key pool, i.e. $\beta_1, \beta_2, \ldots \beta_m$. For each revoked key ring, the MBSs cooperatively generate m new random keys $k'_{\beta_1}, k'_{\beta_2}, \ldots k'_{\beta_m}$.

Subsequently, in connecting to a new MSN, any of the M MBSs distributes the global revocation information by broadcasting a signed list of revoked node identifiers. Subsequently, some sensors at the MSN erase the corresponding revoked keys.

Each sensor u, which has erased any of the revoked keys, contacts the MBS to order a replacement of its revoked keys. Then, the MBS calculates which keys are to be replaced in u (recall this is a simple operation by using the identifier of each node u and the indices $\beta_1, \beta_2, \ldots \beta_m$ of the revoked keys) and distributes securely the corresponding keys to the sensor.

In receiving a key k'_{β_i}, sensor u stores k'_{β_i} in the empty space of its key ring that was left from the corresponding revoked key k_{β_i}. Observe that, because only the keys are changed and not their identifiers, the initial sensor IDs can still be used to find out which shared keys are included in the sensor key ring.

Key Pool Update. From time to time, one of the MBS must report the set-up server[4] on the last keys replaced. The set-up server updates the key pool with the replaced keys. In this manner, when a new node is to be added to the system, it can be bootstrapped with keys, which currently carry previously deployed nodes.

4.4   Coping with Node Replication

In a replication attack, after compromising the keys of a node c, the attacker replicates c's identity, e.g. by configuring a number of malicious nodes with c's key ring.

If the attacker deploys the replicated nodes in the same MSN, the attack is simple to counter by registering each identity location within the MSN, as suggested in [19]. Alternatively, if the replicated nodes reside in different non-interconnected MSNs, the attack is more difficult to counter. There are two options. First, if we assume that MBSs have mutual interconnection while connected to different MSNs, they can cooperatively real-time check for replicated identities in different MSNs. Second, if MBSs have mutual interconnection when they are close to each other but disconnected from MSNs, they can cooperatively check for replicated identities by registering the node identities found in different MSNs after they visited the MSNs.

4.5   Sensor-MBS Communication

The key revocation system requires very secure communication between sensors and MBSs. In this Section, we describe how to secure the communication sensor-MBS and discuss different communication strategies for an MBS to reach the sensors.

---

[4] Assumed to reside in a very secure perimeter and to be accessible in very restricted conditions.

Security. Following a trivial key pre-distribution scheme, for each mobile base station MBS_i, $i = 1, \ldots M$, and each sensor node u_j, $j = 1, \ldots N$, $M \ll N$, a set-up server randomly picks and distributes a pairwise key k_{u_j, MBS_i}.

This scheme enables each sensor node to communicate securely with any of the M MBSs. It is unconditionally secure and uses computationally efficient symmetric key cryptography.

Communication Modes. To avoid attacks from compromised nodes in the distribution of replaced keys direct communication sensor-MBS is required.

In small-scale MSNs (from two to tens of nodes) all the sensors are in wireless transmission range of each other. In this case, direct communication sensor-MBS is feasible and preferred, i.e. the MBS communicates with each and every sensor in the MSN by using a single-hop link. In medium-scale MSNs (hundreds of nodes) sensors connect by using multi-hop links. In this case, the MBS can directly connect to each and every sensor by moving around the MSN deployment space. In large-scale MSNs sensors (thousands of nodes) connect by using multi-hop links. In very sparse networks it may be unfeasible for a single MBS to directly connect to each and every sensor by moving around the MSN deployment space. In this case, the MSN coverage space should be split in different areas $A_1, \ldots A_a$ and responsibility for each area should be assigned to different MBSs, e.g. to $MBS_1, \ldots MBS_a$. In this manner, each MBS_i can directly connect to each and every sensor in its respective MSN area A_i, $i = 1, \ldots a$, by moving around the MSN area A_i.

5   Analysis

In this Section we evaluate the performance of the key revocation and replacement system. This includes assessing if the system meets the goals of keeping KPS initial connectivity and resiliency in an efficient and effective manner. In this paper, we do not evaluate the performance of the underlying IDS or revocation systems (this may be found in the corresponding papers [3,4,19,20,21,22,24]. We assume that the global IDS attack detection ratio r, $0 < r < 1$, can be measured and quantified in different MSN configurations [23]. We use r as input in our analysis.

5.1   KPS Connectivity

To analyze the connectivity of a secure MSN we use Eschenauer and Gligor method [3] based on Erdös and Rényi model [25] (other models could also be employed as suggested in [9]). We analyze the case where MSNs have a minimum number of 1000 nodes and the total number N of system nodes is 10000. In Erdös and Rényi model for large n, $(n > 1000)$, to have an almost connected random graph the required degree d of a node varies insignificantly with n, e.g. for $n = 10000$, d increases only by 2 in respect to d for $n = 1000$ (refer to Figure 1 in [3] for more details). This means that we can calculate the parameters of the KPS for $n = 10000$ and the design will be valid for MSNs of sizes from 1000 to 10000 nodes [3].

The connectivity of a random graph $G(n, p)$ (where n is the total number of sensors and p is the probability that two sensors share at least a key) is restricted by the physical wireless connectivity constraints of the MSN. To achieve the desired connectivity, the probability p'_{req} that two sensors share a key is to be calculated using the expected node degree d and the required size n'_{req} of a node's neighborhood:

$$p'_{req} = \frac{d}{n'_{req} - 1} \ . \tag{1}$$

The actual probability p'_{act} that two sensors share a key is a function of the keys stored in a sensor and the size of the key pool [3]:

$$p'_{act} = 1 - \frac{\left(1 - \frac{k}{|S|}\right)^{2(|S|-k+1/2)}}{\left(1 - \frac{2k}{|S|}\right)^{(|S|-2k+1/2)}} \ . \tag{2}$$

Initially, the KPS is designed to guarantee $p'_{act} > p'_{req}$, while considering sensors memory constraints (i.e. $k \le m$) and trying to maximize resiliency ρ_{rx} against sensor captures, $0 < \rho_{rx} < 1$. In the limit, initially $p'_{act} = p'_{req}$ can be chosen to guarantee desired connectivity and maximize resiliency. But, what happens if each sensor has to progressively erase keys from its key ring? The effect is that p'_{act} decreases to p'_{act_new} (i.e. $p'_{act\_new} < p'_{act}$) and, thus, eventually goes under the initially designed value p'_{req}. Consequently, MSN secure connectivity progressively decreases.

In our system, MBSs monitor the current value p'_{act_new}. The key replacement process should start whenever $p'_{act\_new} < p'_{req} - \varepsilon$, where ε measures the maximum accepted reduction of the initially designed p'_{req}. In this manner, our solution keeps original MSN secure connectivity throughout the whole system lifetime because the number of keys stored in each sensor key ring does not progressively decrease but can be maintained above the lower limit to continuously guarantee $p'_{act\_new} \ge p'_{req} - \varepsilon$.

5.2 KPS Resiliency

The resiliency of the KPS is measured as the fraction of total network communications that are compromised when x nodes are captured [4]. If no IDS is used, then each captured node contributes to increase the fraction of compromised communications. However, if an IDS with attack detection ratio r is used and the detected compromised keys are revoked, then only undetected captured nodes contribute to the fraction of compromised communications.

Let the number of total captured sensors be x. The expected fraction of total keys compromised is $\rho_x = 1 - \left(1 - \frac{m}{|S|}\right)^x$ [4]. Let us assume that each of the x captured sensors runs one attack. Then, exactly $(1 - r)x$ captured sensors go undetected and the expected fraction of the total undetected compromised keys is $\rho_{(1-r)x}$.

The set of different keys carried by the rx detected captured nodes and the set of different keys carried by the $(1 - r)x$ undetected captured nodes have exactly a fraction $\delta_{rx} = \rho_{rx} + \rho_{(1-r)x} - \rho_x$ of keys in common. Then, assuming that the undetected captured nodes order replacement of their revoked keys during the key replacement process, a fraction δ_{rx} of the replaced keys will also be distributed to the undetected captured nodes. Consequently, the undetected captured nodes will still possess a fraction $\rho_{(1-r)x}$ of compromised keys. Otherwise, if undetected captured nodes do not order the replacement of their revoked keys, then the fraction of compromised keys is reduced to $\rho_{(1-r)x} - \delta_{rx}$ after the key replacement process, i.e. resiliency increases.

5.3   Communication Cost

The attacker needs an interval of time I to compromise the keys of x nodes. During I a number rx of nodes is detected to be compromised. Then, the expected number of total keys to be replaced per node is $\rho_{rx} \times m$. The MBS sends to each different sensor the corresponding keys by unicast secure communication. Consequently, the communication overhead per sensor is $\rho_{rx} \times m \times \lfloor \log q \rfloor$ bits.

Assume $m = 200$, $|S| = 10000$ and $\lfloor \log q \rfloor = 64$ bits. The number of bytes exchanged by a sensor and the MBS is under 0.9 kbytes for a wide span of x and r. We assume I to be in the order of hours, days or weeks. Consequently, the communication overhead does not significantly effect network bandwidth (within the sensor neighborhood) or energy consumption.

6   Conclusions

Key management is paramount for MSN security. The special nature of MSNs imposes challenging requirements on key management design.

In this paper, we have reviewed current key management techniques proposed for wireless sensor networks in the context of MSNs and identified open issues for research. Deterministic KPSs show excellent connectivity properties for MSNs but work is still needed to improve their resiliency. Random KPS can also be applied to some MSN applications (when the size of MSNs does not go under certain limits) and presents good resiliency properties. However, their connectivity properties decrease when keys are revoked.

We have also proposed a novel approach to control and replace pool-based pre-distributed keys, which sets a first step towards improving connectivity properties of pool-based KPSs. The analysis in this paper demonstrates that initial connectivity and resiliency properties of the key pre-distribution scheme are maintained even when nodes are revoked.

References

1. R. Blom: An optimal class of symmetric key generation systems. In Proc. of the EUROCRYPT 84 workshop on Advances in cryptology: theory and application of cryptographic techniques. (1985) 335 – 338.

2. C. Blundo, A. De Santis, A. Herzberg, S. Kutten, U. Vaccaro, M. Yung: Perfectly-Secure Key Distribution for Dynamic Conferences. (1992) 471–486.
3. L. Eschenauer and V. D. Gligor: A key-management scheme for distributed sensor networks. ACM Conference on Computer and Communications Security. (2002) 41–47.
4. H. Chan, A. Perrig and D. X. Song: Random Key Predistribution Schemes for Sensor Networks. IEEE Symposium on Security and Privacy. (2003) 197.
5. W. Du, J. Deng, Y. S. Han and P. K. Varshney: A pairwise key pre-distribution scheme for wireless sensor networks. ACM Conference on Computer and Communications Security. (2003) 42–51.
6. D. Liu, P. Ning and R. Li: Establishing pairwise keys in distributed sensor networks. ACM Trans. Inf. Syst. Secur. (2005) 41–77.
7. Hwang and Y. Kim: Revisiting random key pre-distribution schemes for wireless sensor networks. In Proc. of the 2nd ACM SASN workshop. (2004) 43 – 52.
8. S. Zhu, S. Setia and S. Jajodia: LEAP: Efficient Security Mechanisms for Large-Scale Distributed Sensor Networks. ACM CCS Conference. (2003) 62–72.
9. R. Di Pietro, L. V. Mancini and A. Mei, A. Panconesi: Connectivity Properties of Secure Wireless Sensor Networks. In Proc. of the 2nd ACM SASN workshop. (2004) 53 – 58.
10. S. Basagni, K. Herrin, E. Rosti, and D. Bruschi: Secure pebblenets. In Proc. of the 2nd ACM International Symposium on Mobile Ad Hoc Networking and Computing. (2001) 156–163.
11. H. Baldus, K. Klabunde, and G. Muesch: Reliable Set-UP of Medical Body-Sensor Networks. In Proc. of the European Workshop on Wireless Sensor Networks. (2004) 353–363.
12. P. Juang, H. Oki, Y. Wang, M. Martonosi, L. Peh, and D. Rubenstein: Energy-Efficient Computing for Wildlife Tracking: Design Tradeoffs and Early Experiences with Zebranet. In Proc of the Tenth International Conference on Architectural Support for Programming Languages and Operating Systems. (2002).
13. S. A. Camtepe and B. Yener: Combinatorial Design of Key Distribution Mechanisms for Wireless Sensor Networks. In Proc. of Computer Security- ESORICS. (2004) 293–308.
14. J. Lee, and D. R. Stinson: Deterministic Key Predistribution Schemes for Distributed Sensor Networks. In Proc. 11th International Workshop. (2004) 294–307.
15. D. Sánchez and H. Baldus. A Deterministic Pairwise Key Pre-distribution Scheme for Mobile Sensor Networks. In Proc. of the First International Conference on Security and Privacy for Emerging Areas in Communication Networks. (2005) 277–288.
16. H. Chan, A. Perrig and D. Song: Key distribution techniques for sensor networks. Wireless sensor networks. Kluwer Academic Publishers. (2004) 277 – 303.
17. A. Perrig, R. Szewczyk, V. Wen, D. Cullar, and J. D. Tygar: SPINS: Security protocols for sensor networks. In Proc. of MOBICOM. (2001).
18. K. Romer and F. Mattern: The design space of wireless sensor networks. IEEE Wireless Communications, Vol. 11, Issue 6. (2004) 54–61.
19. J. Newsome, E. Shi, D. Song and A. Perrig: The sybil attack in sensor networks: analysis & defenses. In Proc. of the third international symposium on Information processing in sensor networks. (2004) 259 – 268.
20. Y. Zhang and W. Lee: Intrusion Detection in Wireless Ad-Hoc Networks. In Proc. of The 6th International Conference on Mobile Computing and Networking. (2000) 275–283.

21. P. Brutch and C. Ko.: Challenges in intrusion detection for wireless ad-hoc networks. In Proc. of the Symposium on Applications and the Internet Workshops. (2003) 368 – 373.

22. S. Ganeriwal and M. B. Srivastava: Reputation-based Framework for High Integrity Sensor Networks. In Proc. of the 2nd ACM workshop on Security of ad hoc and sensor networks. (2004) 66 – 77.

23. D. Watkins and C. Scott: Methodology for evaluating the effectiveness of intrusion detection in tactical mobile ad-hoc networks. In Proc. of the IEEE Wireless Communications and Networking Conference. (2004) Vol. 1, 622 – 627.

24. K. Nadkarni and A. Mishra A novel intrusion detection approach for wireless ad hoc networks. In Proc. of the IEEE Wireless Communications and Networking Conference. (2004) Vol. 2, 831–836.

25. J. Spencer: The Strange Logic of Random Graphs. Algorithms and combinatorics 22. Springer-Verlag SBN 3-540-41654-4. (2000).

Server-Aided RSA Key Generation Against Collusion Attack

Yun Chen[1,2,*], Reihaneh Safavi-Naini[2], Joonsang Baek[2], and Xin Chen[3]

[1] Chengdu University of Information Technology, PR China
[2] University of Wollongong, Australia
[3] Zhenzhou University of Light Industry, PR China

Abstract. In order to generate RSA keys on low-power hand-held devices, server-aided RSA key generation protocols [2] were proposed. One drawback of these protocols, however, is that they cannot prevent a "collusion attack" in which two key generation servers communicate with each other to get useful information about the user's private key. In this paper, we present two new server-aided RSA key generation protocols secure against such an attack. In addition to this, we adopt a fast primality test in our protocols, which is locally run on a hand-held device. In the concluding section a weakness of the proposed protocol is discussed.

Keywords: Information security, RSA crypto-system, server-aided key generation, collusion attack, hand-held device.

1 Introduction

Over the last decade, a number of new applications for hand-held devices such as PDAs, smart-cards and the PalmPilot have been emerged. For example, wireless purchase using a mobile phone, hand-held electronic wallets and using a handheld device as an authentication token are becoming common in everyday life. On the other hand, most of these applications require secure means of communication such as encryption or digital signature schemes to perform their tasks without compromising confidentiality or authenticity. For most of practical applications, keys for small devices are generated by a trusted server prior to deployment of these small devices, it is convenient for a handheld user to generate his own keys and use the keys to freely access in or out mobile ad-hoc network communications. Currently, RSA is the most prevailing cryptosystem for realizing such schemes.

As well known, to implement the RSA cryptosystem, one needs to generate large primes. However, this causes a large amount of computations due to a computationally-heavy primality test algorithm. Especially for low-power hand-held devices that only have limitative computational resources, this burden may be even heavier. As an example, it was reported that the generation of a 1024-bit RSA key takes about 15 minutes on the PalmPilot [2].

* Financially supported by National Laboratory for Modern Communications project, PR China (Code:51436010404DZ0235).

M. Burmester and A. Yasinsac (Eds.): MADNES 2005, LNCS 4074, pp. 27–37, 2006.

An immediate solution for the above problem one might consider is to have a server generate the key and send it back to the hand-held devices. However, the problem with this approach is that the server needs to be trusted unconditionally as it learns the user's key. N. Modadugu, D. Boneh and M. Kim [2] resolved the problem by presenting server-aided RSA key generation protocols, which we call "MBK". Using any of the MBK protocols, users can quickly generate an RSA key on a hand-held device with the help of *untrusted* servers in such a way that once the key is generated the server should not get any information about the key it helped generation.

However, one drawback of the MBK protocols is that two servers involved in the generation of RSA keys *must not* share information with each other. Otherwise, they can easily acquire the user's secret key as noted in [2].

For example, an e-shop's internet service provider (ISP) provides two different servers for the purpose of helping clients to generate their private keys. If it is a malicious one, the ISP could easily make the two servers communicate with one another. For another instance, if an eavesdropper has a sniffer connected to a client, the eavesdropper would be able to intercept all of the messages sent by the client.

In this paper, we present two new version of the MBK protocols that resists a "collusion attack" in which the two servers in the server-aided RSA key generation collude to share information. Our solutions are almost as efficient as the original MBK protocols and hence can be widely used in practice. In section 6 we outline an attack on the protocol that became known to authors after the presentation of the paper in the conference.

2   Overview of RSA Key Generation

In this section we give a brief overview of the RSA key generation.

As a result of the RSA key generation, a key $\langle N, e, d \rangle$ can be generated. Here $N = pq$ denotes an n-bit modulus, where p and q are two large primes, each of which is $\frac{n}{2}$ bits long. e denotes the public exponent and d denotes the private exponent. A high-level description of the RSA key generation algorithm is as follows [2].

1. Generate two primes p and q through the following steps:
   (a) Pick a random $\frac{n}{2}$-bit candidate value p or q.
   (b) Check that p is not divisible by any small primes (i.e. 2,3,5,7, etc.) using trial division.
   (c) Pick $g \in [2, p-1]$ or $g \in [2, q-1]$ at random and check whether

$$g^{\frac{p-1}{2}} \equiv \pm 1 \ (\text{mod } p) \ \ or \ \ g^{\frac{q-1}{2}} \equiv \pm 1 \ (\text{mod } q) \tag{1}$$

   is satisfied. If Equation (1) holds on every random number g, p or q is considered to be a pseudo-prime(Note that all primes will pass this test while composites will fail with overwhelming probability).
2. Compute $N = pq$.

3. Pick encryption and decryption exponents e and d such that $ed \equiv 1 \pmod{\varphi(N)}$, where $\varphi(N) = (p-1)(q-1)$. (Note that φ denotes the Euler function).

In the above algorithm, the most resource-consuming operations occur in Step 1(c), which is the primality testing part. Compared with 1(c), computational overhead for conducting Step 2 and Step 3 are small.

We remark that when RSA cryptosystem is used to encrypt a short session key (for a symmetry encryption scheme), which is shorter than the bit-length of p, one can use the "unbalanced RSA", firstly considered by Shamir [3]. In his unbalanced RSA, the encryption and decryption exponents e and d satisfy $ed \equiv 1 \pmod{(p-1)}$ and the size of a prime factor q is approximately 10 times larger than the other prime factor p. Shamir [3] noted that the key generation of the standard RSA can be sped up in the unbalanced version without extra time complexity. In [2], N. Modadugu et al. present a variant of the unbalanced RSA where a modulus is a product of a prime and a certain size of random number. It is evident that in this variant, the computations in Step 1(c) can be greatly reduced due to the fact that only one prime needs to be generated. (As a result, the RSA key generation in [1] can be sped up further).

3   MBK Protocols

In this section we review the MBK protocols for standard and unbalanced RSA key generation.

3.1   Standard RSA Key Generation

The MBK protocol to generate a standard RSA modulus can be described as follows:

1. Hand-held generates two candidates p and q such that $p \equiv q \equiv 3 \pmod 4$ and none of them is divisible by small primes.
2. Hand-held computes $N = pq$ and $\varphi(N) = N - p - q + 1$. It then picks $g \in Z_N^*$ at random.
3. Hand-held picks $t_1, t_2 \in [-N, N]$ at random such that $t_1 + t_2 = \frac{\varphi(N)}{4}$.
4. Hand-held sends $\langle N, g, t_1 \rangle$ to Server 1 and $\langle N, g, -t_2 \rangle$ to Server 2.
5. Server 1 computes $X_1 \equiv g^{t_1} \pmod N$. Server 2 computes $X_2 \equiv g^{-t_2} \pmod N$. Both results X_1 and X_2 are sent back to Hand-held.
6. Hand-held checks if $X_1 \equiv \pm X_2 \pmod N$. If so, $N = pq$ is declared as a potential RSA modulus. Otherwise, the algorithm is restarted from Step 1.
7. Hand-held locally runs a probabilistic primality test to verify that p and q are primes. This is done to ensure that the servers returned correct values.

It is remarkable that when generating an n-bit RSA key, a single primality test takes little time compared to the search for an $(\frac{n}{2})$-bit prime. So, step 7 adds very little to the total running time. The verification of soundness of the above protocol can be found in [2].

In the above protocol, most expensive computations are offloaded onto the servers. Although many candidates are generated during the search for RSA modulus, both Server 1 and Server 2 learn no extra information from rejected candidates as these candidates are independent of each other. Notice that once $N = pq$ is fixed, it becomes public. Hence, any information from N cannot help the servers to get private information of the Hand-held. Also, since t_1 and t_2 are chosen at random, they provide no extra information for Server 1 and Server 2 to attack RSA assuming that there is no communication between the two servers.

It is obvious that the MBK protocol is insecure against collusion attack as if the two servers communicate, they can compute $\varphi(N) = 4(t_1 + t_2)$, which leads to break RSA cryptosystem.

3.2 Unbalanced RSA Key Generation

Now, we describe the MBK protocol to generate an unbalanced RSA modulus as follows:

1. Hand-held generates an $\frac{n}{2}$-bit candidate p that is not divisible by small primes and random number R which is 8-10 times large in size than p. We require that $p \equiv 3 \pmod{4}$.
2. Hand-held computes $N = pR$.
3. Hand-held picks $t_1, t_2 \in [-N, N]$ such that $t_1 + t_2 = \frac{p-1}{2}$ at random. It then picks $g \in Z_N^*$ at random.
4. Hand-held sends $\langle N, g, t_1 \rangle$ to Server 1 and $\langle N, g, -t_2 \rangle$ to Server 2.
5. Server 1 computes $X_1 \equiv g^{t_1} \pmod{N}$. Server 2 computes $X_2 \equiv g^{-t_2} \pmod{N}$. Both results X_1 and X_2 are sent back to Hand-held.
6. Hand-held checks if $X_1 \equiv \pm X_2 \pmod{p}$. If so, $N = pR$ is declared as a potential unbalanced RSA modulus. Otherwise, it restarts the algorithm from Step 1.
7. Hand-held locally runs a probabilistic primality test to verify that p is a prime.

In the above protocol, the total running time is reduced by a factor of 5 [2]. Here, it is worth emphasizing that an unbalanced RSA is only suitable for the case when the size of a plain-text is smaller than that of prime p. Besides, the unbalanced RSA key can only be used for encryption. It cannot be used for signature.

4 Proposed Solutions: Two New RSA Server-Aided Key Generation Protocols Secure against Collusion Attack

Recall that in MBK protocol, if two key generation servers collude, it is easy for them to obtain the hand-held device's private key. More precisely, if the two servers exchange t_1 and t_2 in the standard RSA version of the MBK protocol, they can easily recover $\varphi(N)$ by simply computing $\varphi(N) = 4 \times (t_1 + t_2)$. Furthermore, they can figure out the user's private key d from $\varphi(N)$ and public key e by running the Euclidean algorithm [5] to get $d \equiv e^{-1} \pmod{\varphi(N)}$.

In this section, we present two new RSA server-aided key generation protocols secure against the above collusion attack.

4.1 Standard RSA Key Generation Against Collusion Attack

A standard RSA key generation version of the new protocol can be described as follows:

1. Hand-held generates two candidates p and q so that each of them is $\frac{n}{2}$-bit in length; and none of them is divisible by small primes.
2. Hand-held computes $N = pq$ and $\varphi(N) = N - p - q + 1$.
3. Hand-held picks a random $g \in Z_N^*$ such that g is not a multiple of p nor q.
4. Hand-held picks at random a large odd integer k in $[-N, N]$ such that $\gcd(k, \varphi(N)) = 1$. It then picks integers t_1 and t_2 at random in the range $[-N^2, N^2]$ such that $t_1 + t_2 = k\varphi(N) = T$, where none of t_1, t_2 and T is multiple of either p or q.
5. Hand-held sends $\langle N, g, t_1 \rangle$ to Server 1 and $\langle N, g, t_2 \rangle$ to Server 2
6. Server 1 computes $X_1 \equiv g^{t_1} (\bmod N)$. Server 2 computes $X_2 \equiv g^{t_2} (\bmod N)$. All of the results are sent back to Hand-held.
7. Hand-held checks if $X_1 X_2 \equiv 1 (\bmod N)$. If so, p and q are considered to be potential primes. Then, $N = pq$ is declared as a potential RSA modulus where $\varphi(N)$ must be equal to $(p-1)(q-1)$. Otherwise, it restarts the operations from Step 1.
8. Hand-held selects a small Mersenne prime m and locally runs a probabilistic primality test to check whether $m^{\frac{p-1}{2}} \equiv \pm 1 (\bmod p)$ and $m^{\frac{q-1}{2}} \equiv \pm 1 (\bmod q)$ hold or not. If the candidates pass the test, we can confirm that p and q are real primes. Otherwise, it restarts the operation from Step 1.
9. Compute $N = pq$.
10. Pick encryption and decryption exponents e and d such that $ed \equiv 1 \pmod{\varphi(N)}$, where $\gcd(e, T) \neq 1$.

Now, we verify the soundness of the above protocol.

Theorem 1. *Suppose $\gcd(g, N) = 1$. If both p and q are primes, $X_1 X_2 \equiv 1 (\bmod N)$ will hold in Step 7. Whereas if one of p and q is composite, $X_1 X_2 \equiv 1 (\bmod N)$ will fail with overwhelming probability.*

Proof. Suppose that p and q are primes. If Step 3 is performed correctly, then $\gcd(g, N) = 1$, we hereby have

$$g^{\varphi(N)} \equiv 1 \pmod{N} \tag{2}$$

Now, raise both sides of Equation (2) to the power of k chosen by Hand-held in Step 4. We then have

$$g^{k\varphi(N)} \equiv 1^k \equiv 1 \pmod{N}. \tag{3}$$

Since $k\varphi(N) = t_1 + t_1$, we have $g^{k\varphi(N)} = g^{t_1 + t_2} = X_1 X_2$. Combining this with (3), we obtain

$$X_1 X_2 \equiv 1 \pmod{N}. \tag{4}$$

On the other hand, if one of p and q is not prime, Equation (2) will fail with overwhelming probability. Consequently, Equations (3) and (4) will fail with overwhelming probability.   □

Theorem 2. *Suppose $gcd(g, N) = 1$. If $X_1 X_2 \equiv 1 (\bmod \ N)$ is true in Step 7, $g^{\varphi(N)} \equiv 1 (\bmod \ N)$ must hold.*

Proof. Raise both sides of Equation (4) to the power of k^{-1}. Since $gcd(k, \varphi(N)) = 1$, we have $kk^{-1} \equiv 1 (\bmod \ \varphi(N))$. Therefore,

$$(X_1 X_2)^{k^{-1}} \equiv g^{k\varphi(N)k^{-1}} \equiv g^{\varphi(N)} \equiv 1 \quad (\bmod \ N) \qquad (5)$$

□

4.2   Unbalanced RSA Key Generation Against Collusion Attack

We now describe an unbalanced version of our server-aided RSA key generation protocol secure against collusion attack.

1. Hand-held generates a $\frac{n}{2}$-bit candidate p which is indivisible by small primes and a random number R which is 8-10 times in size than p. We require that $p \equiv 3 (\bmod 4)$. [2]
2. Hand-held computes $N = pR$.
3. Hand-held picks at random a large odd integer k from the range $[-p, p]$ such that $gcd(k, (p-1)) = 1$. It then picks integers t_1 and t_2 at random in the range $[-N, N]$ such that $t_1 + t_2 = k(p-1) = T$,
4. Hand-held picks a random number $g \in Z_N^*$ such that g is not multiple of p
5. Hand-held sends $\langle N, g, t_1 \rangle$ to Server 1 and $\langle N, g, t_2 \rangle$ to Server 2
6. Server 1 computes $X_1 \equiv g^{t_1} (\bmod \ N)$. Server 2 computes $X_2 \equiv g^{t_2} (\bmod \ N)$. Both results X_1 and X_2 are sent back to Hand-held.
7. Hand-held checks if $X_1 X_2 \equiv 1 (\bmod \ p)$. If so, p is considered to be potential prime. Then, $N = pR$ is declared as a potential RSA modulus where $\varphi(p)$ must be equal to $(p-1)$. Otherwise, it restarts the operations from Step 1.
8. Hand-held chooses a small Mersenne prime m and locally runs a probabilistic primality test. If $m^{\frac{p-1}{2}} \equiv \pm 1 (\bmod \ p)$, then p is a real prime.
9. Compute $N = pR$.
10. Pick encryption and decryption exponents e and d such that $ed \equiv 1 (\bmod \ (p-1))$, where $gcd(e, T) \neq 1$.

It is easy to verify the soundness of the above protocol. Note that if p is prime, we should have

$$g^{p-1} \equiv 1 \quad (\bmod \ p) \qquad (6)$$

It follows that

$$g^{k(p-1)} = g^{t_1 + t_2} \equiv X_1 X_2 \equiv 1 \quad (\bmod \ p) \qquad (7)$$

for k chosen by Hand-held in Step 3.

On the other hand, if p is not prime, Equation (6) will fail with overwhelming probability and so does Equation (7).

We formally state this as the following theorem.

Theorem 3. *Suppose $gcd(g, (p - 1)) = 1$. If p is prime, $X_1X_2 \equiv 1(\mathrm{mod}\ p)$ will hold in Step 7. Whereas if p is composite, $X_1X_2 \equiv 1(\mathrm{mod}\ p)$ will fail with overwhelming probability.*

Note that in the proof of the above theorem, we have

$$(X_1X_2)^{k^{-1}} \equiv g^{p-1} \equiv 1 \quad (\mathrm{mod}\ p) \tag{8}$$

where $kk^{-1} \equiv 1(\mathrm{mod}\ (p - 1))$. Hence we obtain the following theorem.

Theorem 4. *Suppose $gcd(g, (p-1)) = 1$. If $X_1X_2 \equiv 1(\mathrm{mod}\ p)$ (Equation (7)), $g^{p-1} \equiv 1(\mathrm{mod}\ p)$.*

5   Analysis of Our Protocols

5.1   Performance Analysis

In Step 8 of the two proposed protocols, we use a small Mersenne prime to perform the primality test. The reasons for this are the following: 1) Small Mersenne prime will satisfy $gcd(m, N) = 1$. 2) Multiplication by m can be substituted with a simple shift and subtraction operations.

We now show that due to the introduction of the small Mersenne prime, our RSA key generation protocol presented in Section 4.1 performs better than the same kind of the MBK protocol. (The same result applies to the unbalanced RSA key generation protocol too).

We note that assuming that the modular exponentiation is computed by the Square-And-Multiplication (SAM) algorithm [8] and the exponent is balanced by 0's and 1's on average, the number of steps for modular multiplication is $\frac{1}{3}$ of the total number of recursive steps in the SAM algorithm [7]. We also note that small devices generally use the byte-wise operation. The main difference of the performance between MBK protocols and our protocols lies in the number of byte-wise multiplications and additions (or subtractions) in one modular multiplication, which are listed in Table 1. (In this table, n denotes the binary length of N; n_{m1} and n_{m2} denote the number of byte-wise multiplications needed in MBK and in our protocol respectively; n_{a1} and n_{a2} denote the number of byte-wise additions or subtractions needed in MBK protocol and our protocol respectively).

Table 1. Number of byte-wise computations

n	n_{m1}	n_{m2}	n_{a1}	n_{a2}
1024	4096	0	8191	≤ 255
2048	16384	0	32767	≤ 511

Table 1 shows that the number of byte-wise computations for one modular multiplication in our protocols is much smaller than that in the MBK protocols.

Since about $\frac{n}{4}$ modular multiplications are needed in a modular exponentiation [7], the improvement on performance in our protocol is considerable.

Although the workload on servers is doubled in our protocol, it is acceptable as the server is usually assumed to have enough computational resources. We compiled the primality test program using Visual C++ 6.0, where traditional SAM algorithm is used. The average running time for generating a 512 bits prime on Pentium 4 with 2.0GHz and 256M RAM is 39 seconds. It can be reduced below 3 seconds when the Sliding Window and the Montgomery algorithms are added to SAM algorithm.

5.2   Security Analysis

In this section, we analyze the security of our server-aided RSA key generation protocols.

According to the analysis given in [2], the untrusted servers learn no useful information about RSA private key: During the search for the RSA modulus, many candidates are generated. Since these candidates are independent of one another, any information from the rejected candidates does not help the servers to attack the final RSA modulus N. Once N is decided, it will become public for it is part of public key. Hence it discloses no extra information. In the MBK protocol, t_1 is just a random value for Server 1. It provides no useful information to Server 1 for assisting the attacks on RSA. So is the t_2 for Server 2. The assumption, however, is that there is no communication between the two servers, namely, they do not collude.

Also in our protocol, the untrusted servers shall not get any useful information even if they share information with each other. Formally, we state and prove the following theorem.

Theorem 5. *Our server-aided key generation protocol for standard RSA is secure against collusion attack.*

Proof. Suppose that in our protocol, Servers 1 and 2 have communicated with each other and have derived

$$T = k\varphi(N) \tag{9}$$

from their sharing information.

We argue that it is computationally infeasible for the servers to get $\varphi(N)$ from T. If k is a large odd integer chosen at random from $[-N, N]$, the bit length of T will be far greater than 1024. In fact, this is an integer factorization problem: It is known that using the general number field sieve method, RSA modulus of a bit-length 576 has been factored. [4].

Now, we assume that the two servers have luckily picked a random integer S which is a multiple of $\varphi(N)$, from the range $[-N^2, N^2]$. (S and T are similar in size). Let $S = l\varphi(N)$, where l is some integer. In this case, $\varphi(N)$ is a common divisor of S and T. Using the Euclidean algorithm, the servers can compute a greatest common divisor G of S and T, namely

$$G = \gcd(S, T) \tag{10}$$

It is obvious that G must contain a factor $\varphi(N)$ and is much smaller than S or T. So the adversaries might work out $\varphi(N)$ from G. In a special case when l is relatively prime to T, $i.e.$ $\gcd(l,T) = 1$, the servers can get

$$G = \varphi(N) \tag{11}$$

Using $\varphi(N)$, the servers would be able to get user's private RSA key. However, the success probability that a randomly chosen integer is a multiple of $\varphi(N)$ is at most

$$P_s = \frac{k}{2N^2}. \tag{12}$$

Assuming that the size of k is similar to that of N, say 1024 bits, we have $P_s \leq \frac{1}{2^{1025}}$, which is negligible.

A similar situation will occur when a randomly chosen integer is a multiple of $(p-1)$ or $(q-1)$, say $s_1 = x(p-1)$ or $s_2 = y(q-1)$. This time, the success probability with respect to s_1 will be

$$P_{s_1} = \frac{k(q-1)}{2N^2} \tag{13}$$

and the success probability with respect to s_2 will be

$$P_{s_2} = \frac{k(p-1)}{2N^2} \tag{14}$$

Assuming that the size of k is similar to that of N, we obtain $P_{s_1} \leq \frac{1}{2^{513}}$ and $P_{s_2} \leq \frac{1}{2^{513}}$, which are also negligible. Therefore, it is computationally infeasible for the servers to acquire user's private information by random guessing.   □

There is another potential way to break RSA. Suppose $\gcd(e, T) = 1$, then some t and r must exist such that

$$et + rT = et + rk\varphi(N) = 1 \tag{15}$$

Equation (15) can be rewritten as

$$et \equiv 1 \;(\mathrm{mod}\ \varphi(N)) \tag{16}$$

Hence, we must have

$$t = d \tag{17}$$

It is the reason we require that $\gcd(e, T) \neq 1$ at step 10 in our protocols.

The above analysis implies that the two untrusted servers know nothing about the user's RSA private key even if they can share information with each other.

We remark that due to the characteristic of resisting collusion attack, Handheld is also able to generate RSA modulus with help of single server.

With a similar argument, the unbalanced RSA version of our protocol can be proven to be secure against collusion attack. We state the following theorem.

Theorem 6. *Our server-aided key generation protocol for unbalanced RSA is secure against collusion attack.*

First of all, servers can not derive $\varphi(N)$ from factorizing T. Secondly, for a randomly chosen integer S in $[-N^2, N^2]$, its probability of being $(p-1)$- multiple is at most $\frac{1}{2R} = \frac{1}{2^{4097}}$, which is certainly negligible, where R is 4096 bits long.

Next, we should point out that there is a potential way for malicious servers to acquire user's unbalanced RSA private key.

Suppose r is a random variable, Let $V = T + r$. If k is not big enough, the colluding servers are able to exhaustively search for r. Once $r = k$, then the servers get p from $gcd(V, N) = gcd[k(p-1)+r, pR] = p$, as long as $gcd(k, R) = 1$. The countermeasure against exhaustive search attack is to choose large enough k.

Likewise, the two servers cannot derive d from e and T so long as $gcd(e, T) \neq 1$.

From the analysis above we know that a Hand-held is able to protect its unbalanced RSA private key from collusion attack as long as k is big enough.

6   An Attack and Concluding Remarks

After the workshop it was brought to our attention that the following attack on the protocol allows the exponent to be found:

"The protocol requires $gcd(e, T) = a > 1$ as otherwise the decryption exponent, d, can be easily computed. However, $gcd(e, T) = a > 1$ means that $a|e$ and $a|k\phi(n)$, and since $gcd(e, \phi(n)) = 1$, we have $a|k$. Letting $T^* = T/a$ and $k^* = k/a$, we get $T^* = k^*\phi(n)$ and hence $gcd(T^*, e) = 1$."

Our early investigation showed that unfortunately the attack cannot be escaped by an easy fix and so the protocol as described above is insecure.

The MBK protocols [2] efficiently solved the problem of RSA key generation on low power hand-held devices. However, it turns out that these protocols cannot prevent collusion attack [2]. Our proposed solution to this problem is insecure and so the problem of finding a protocol that resist collusion attack, remains open.

Acknowledgement. The first author would like to thank Dr. Luck McAven and Dr. Yi Mu for their assistance on formatting this article, and Wei Zhang for compiling a primality test program.

References

1. D. Boneh and M. Franklin, "Efficient Generation of Shared RSA Keys", Journal of the ACM, 48(4):702–722, 2001.
2. N. Modadugu, D. Boneh, and M. Kim, "Generating RSA Keys on a Handheld Using an Untrusted Server," In CT-RSA 2000, Available at http://crypto.stanford.edu/ dabo/pubs.html.
3. A. Shamir, "RSA for Paranoids", CryptoBytes, 1(3):1-4, 1995.
4. RSA Security, "RSA-576 is Factored!", Available at http://www.rsasecurity.com/rsalabs.

5. H. Cohen, "A Course in Computational Algebraic Number Theory", Springer-Verlag:Vol.138 of Graduate Texts in Mathematics, 1996
6. G. Horng, "A Secure Server-aided RSA Signature Computation Protocol for Smart Cards", Journal of Information Science and Engineering, 16(6):847-855, 2003
7. Y.Chen, and Y.H.Gong, "An Improved Algorithm for Recursive Sums of Residue", Journal of UEST (Chinese),29(1):1-4,2001
8. G.Gong, and L.Harn, "Public-key Cryptosystems Based on Cubic Finite Field Extension", IEEE Transactions on Information Theory, 45(7):2601-2605, 1999

Hybrid Approach for Secure Mobile Agent Computations

J. Todd McDonald*

Florida State University, Tallahassee FL 32306-4530, USA
mcdonald@cs.fsu.edu
http://cs.fsu.edu/~

Abstract. Mobile agent applications are particularly vulnerable to malicious parties and thus require more stringent security measures–benefiting greatly from schemes where cryptographic protocols are utilized. We review and analyze methods proposed for securing agent operations in the face of passive and active adversaries by means of secure multi-party computations. We examine the strengths and weaknesses of such techniques and pose hybrid schemes which reduce communication overhead and maintain flexibility in the application of particular protocols.

1 Introduction

Mobile agents offer a unique method for implementing distributed applications. Itinerant agents have the ability to migrate among a preplanned or ad-hoc set of hosts where host inputs are gathered and agent code is executed. The agent carries both its static code and a dynamic data state which embodies all previous results of execution. Security concerns still occupy a large portion of the research effort associated with such mobile programs–both with protecting agents from malicious hosts and protecting hosts from malicious agents.

A multitude of schemes have been developed for mobile agent security and reviews of various mechanisms have been recorded by McDonald et al. (2005), Bierman and Cloete (2002), and Jansen and Karygiannis (2000). Much work has been done over the last few years to apply the field of theoretical cryptography to the mobile agent security problem. By integrating cryptographic protocols based on secure multi-party computations (SMC), software-only protection mechanism can be designed to guarantee the execution integrity and data confidentiality of an agent while it is executed at a remote host.

The use of secure computation involves a trade-off between security, trust, and overhead. SMC protocols can have varying security attributes–whether at the information theoretic or computational level–and varying levels of communicational and computational overhead–normally considered unreasonable for

---

* The views expressed in this article are those of the author and do not reflect the official policy or position of the United States Air Force, Department of Defense, or the U.S. Government.

M. Burmester and A. Yasinsac (Eds.): MADNES 2005, LNCS 4074, pp. 38–53, 2006.

practical applications. In this paper, we review specifically the use of these approaches for mobile agent security and pose hybrid approaches that offer greater efficiency and more flexibility in integrating SMC protocols.

We organize the paper as follows. Section 2 reviews literature related to secure computations while Section 3 analyzes various efforts to integrate SMC as an agent protection mechanism. Section 4 poses several hybrid mobile agent approaches that minimize communication overhead and add more flexibility in the application of SMC protocols to security. Section 5 summarizes our contributions.

2   Secure Computations

Cryptographers have for some time sought how to perform a group function when there are a number of mutually or partially distrusting participants to the operation. Yao's blind millionaire problem (1986) is often cited as an early formulation for the two-party case where a function $z = f(x, y)$ is computed between Alice and Bob–without leaking any information about Alice's input x or Bob's input y other than what can be deduced from z itself. Goldreich and his colleagues (1987) extend secure computation to n parties–defined in the general case as a publicly available function f that takes n private inputs and returns n private outputs: $f(x_1, x_2, x_3, \ldots, x_n) = (y_1, y_2, \ldots, y_n)$. In some instances, all parties learn the same function output such that $y_1 = y_2 = \ldots = y_n$, making the output publicly known.

Secure computation is referred to synonymously as secure multi-party computation (SMC), secure function evaluation (SFE) or secure circuit evaluation. In terms of practical use, Rabin and Ben-Or (1989) summarize privacy-preserving, real-world applications that can be represented as an SMC problem such as database query, scientific computations, intrusion detection, statistical analysis, geometric computations, and data mining. Malkhi et al. (2004) have developed a full programmatic implementation of a two-party secure function evaluator called Fairplay that uses oblivious transfer [Kilian (1988), Abadi et al. (1989), Abadi and Feigenbaum (1990), Bellare and Micali (1990)] and one-pass Boolean circuits [Yao (1986), Goldreich et al. (1987), Naor et al. (1999), Bellare et al. (1990)].

SMC protocols typically involve several rounds of interaction between parties and assume different types of communication channels including, for example, private channels between every two parties [Goldreich et al. (1987), Chaum et al. (1988), Rabin and Ben-Or (1989)], a broadcast channel [Bellare et al. (1990), Rabin and Ben-Or (1989)], and broadcast subsets among player triples [Fitzi et al. (2005)]. In terms of security, Bellare et al. (1990) recall that the correctness and privacy of any protocol can be reduced to the evaluation of a secure function protocol. In the ideal setting, all parties to an SMC can send their inputs via a secure private channel to a trusted third party that computes the group function and return results fairly.

Naor and Nisim (2001) state a primary security result that any function computable with polynomial resources (communication and computation) can be transformed and computed in a secure manner using polynomial resources. Corruption in multi-party computations deal either with an honest-but-curious (semi-honest) adversary that passively reads information from corrupted parties or an active (malicious) adversary that exerts full control over parties. Privacy of inputs is at issue in passive attacks while correctness of the outputs is more in view in active attacks. Goldreich (2000) concludes that two parties acting maliciously can be forced to behave in a semi-honest manner or else be caught violating the security of the computation.

For any arbitrary function in the presence of an active adversary, Goldreich et al. (1987) conclude the computation can still be securely accomplished as long as less than $\frac{1}{2}$ of the players have not been corrupted. For unconditional security, Ben-Or et al. (1988) and Chaum et al. (1988) say that computations can occur as long as less than $\frac{1}{3}$ of the players have been corrupted and secure channels exist in both directions between any two players. When broadcast channels are introduced, unconditional security is possible for the computation as long as less than $\frac{1}{2}$ of the players are corrupt. Cachin and colleagues [2000] reiterate that computation between two unbounded parties with "full information" is not securely possible for arbitrary functions and only limited to trivial functions g where $g(x, y)$ reveals y. These results are significant when the multi-party computations are applied in the realm of mobile agents.

2.1 Evaluation Techniques and Primitives

Yao (1986) first posed the idea that a function f can be modeled and securely executed as a Boolean circuit in a protocol known as secure circuit evaluation. The circuit can be "scrambled" in a way to secure host inputs and compute the group output. Abadi and Feigenbaum (1990) posed a two-player scheme where one player runs a secret program for another player who has a secret input. Other techniques for circuit construction including multi-party cases have been posed by Goldreich et al. (1987), Ben-Or et al. (1988), Bellare et al. (1990), Naor et al. (1999), and Chaum et al. (1988). Once the function f is represented as a circuit, parties must run a protocol to evaluate every gate in the circuit.

To accomplish secure circuit evaluation, the original wire signals for both inputs and outputs of the circuit are encrypted (garbled) and the actual wire signals used by the parties no longer have their same semantic meaning. As Bellare and Micali (1990) describe, inputs and outputs are translated to their true semantic meaning by two parties by exchanging data in an oblivious manner– typically *1-out-of-2* OT. Naor and Pinkas (2001,2000) have refined OT to achieve greater efficiency and to work in distributed settings.

While OT deals with privacy in circuit-based SMC, cheating can be addressed by verifiable secret sharing (VSS) which allows a "dealer" to distribute shares of a piece of data among different parties [Zhong and Yang (2003), Rabin and Ben-Or (1989), Shamir (1979)]. Normally, parties in the computation must commit to their bits (which become garbled for purposes of evaluation) before they are used. However, no other party could tell whether the scrambled bits actually represent the real semantic meaning of a party's input. By using sharing techniques, parties give shares of their inputs so that any attempt to alter a commitment can be detected. Re-sharing of data to prevent a super adversary with control over some set of parties from gathering enough shares to compromise a system is expounded by Endsuleit and Mie (2003) based on Ostravsky and Yung's technique (1991).

Typically, SMC protocols have been adapted for synchronous networks and suffer from computational or communicational complexity too high for use in the real world. Mobile agents operate in asynchronous environments and therefore other factors must be taken into account before SMC techniques can be applied successfully. Ben-Or et al. (1993, 1994) offer a framework for realistic network environments while Canetti (2001, 2000) has characterized the composable nature of security properties for different protocols operating across asynchronous networks. As Endsuleit with Mie (2003) and Wagner (2004) suggest, timeouts have to be integrated with distributed computations for asynchronous networks (that model the Internet) and the environment for mobile agent applications.

2.2   Single Round Computations

Mobile agents exhibit three unique properties that make using SMC protocols difficult: autonomy, mobility, and disconnected operations. All of the protocols mentioned thus far have relied on the exchange of information between parties in multiple rounds, including the originator of a function. Agents require non-interactive protocols because the originator of a function may be off-line during the actual computation. Autonomy stipulates that the agent does not return home after the first host and can visit some set of known or unknown hosts. Mobility without the help of a trusted third party and minimal communication among parties is a primary goal of agent security schemes. As Abadi and Feigenbaum (1990) and Rivest et al. (1978) discuss, there are two ways to view single round computations between two parties in contrast to traditional secure function evaluation: *computing with encrypted data* and *computing with encrypted functions*. Table 1 summarizes these approaches.

CEF represents the mobile agent transaction scheme best and can be extended easily to a multiple host approach. Sander and Tschudin (1998) posed one of

Table 1. Methods of single-round secure function evaluation

Type	Computation
Computing w/ Encrypted Data (CED)	Alice has input x while Bob holds function $f(\cdot)$. Alice sends an encrypted version of x to Bob who computes and sends the result back to Alice in a single round of interaction. Alice decrypts the result to get $f(x)$ while Bob does not learn x.
Computing w/ Encrypted Functions (CEF)	Alice holds the function $f(\cdot)$ while Bob holds input y. In one-round, Alice sends to Bob an encrypted version of $f(\cdot)$ who provides his input y. Alice receives back and decrypts Bob's result to learn $f(y)$ but does not learn y; Bob does not learn $f(\cdot)$
Secure Function Evaluation (SFE)	Alice and Bob have private inputs to the function $f(x,y)$. Alice and Bob jointly compute the function $f(x,y)$ in one round of computation. Alice learns only the result (and nothing more) while Bob learns neither the result nor Alice's private input.

the first non-interactive CEF approaches for mobile code execution based on homomorphic encryption. Their results were extended to include any function implemented by logarithmic-size circuits. Cachin et al. (2000) developed a non-interactive protocol (which we will refer to as the *CCKM* scheme) that could be used to evaluate all polynomial time functions via the use of a scrambled circuits and oblivious transfer.

Several important results were derived from Cachin et al.: 1) for unbounded passive adversary, any function computable by a polynomial-size circuit can be computed securely; 2) for a bounded active adversary, any function computable by a polynomial-size circuit can be computed securely, given a public key framework; and 3) any function computable by a polynomial-sized circuit has a one-round secure computation in the model.

3   Integrating SMC with Agents

Mobile agent applications have brought a practical relevance to development of secure, efficient cryptographic protocol schemes. The goal of SMC has been stated as guaranteeing the correctness of a function and the privacy of results among the parties. In mobile code systems, similar notions exist: malicious hosts can spy on the code, state, or results of mobile agents that they execute. There are two primary approaches to integrating SMC protocols with mobile agents: *use single agents that implement single-round non-interactive protocols* or *use multiple agents that execute multi-round SMC protocols in coalition schemes*. We discuss approaches and issues with the former next.

3.1   Non-interactive Approaches

Using the notation of Cachin et al. (2000) and Algesheimer et al. (2001), a single-round secure multi-party computation is sent by an agent originator O that embodies a private function to be executed by a set of hosts $H_1, \ldots, H_n$. Two functions describe the computation of an agent in terms of a resulting state $x \in X$ and a current host input $z \in Z$: $x_j = g_j(x_{j-1}, y_j)$ and $z_j = h_j(x_{j-1}, y_j)$. The state update function $g_j(\cdot)$ takes a current state from the set of states X (brought by an agent from the previous host H_{j-1}) and the local host input (taken from the set of all possible inputs Z) and produces a next state x_j. The host output function $h_j(\cdot)$ takes the same inputs as $g_j(\cdot)$ but produces its own local output z_j.

In the CCKM protocol, once the agent computation is represented as a Boolean circuit and encrypted, translation tables are required to map actual signals to scrambled signals. The circuit encoding is based on Yao's (1986) two-party SFE protocol. In order to know what signals to use for their local input, a host performs oblivious transfer with the originator to get a set of scrambled signals, and the originator does not know which signals are chosen. The following security properties are thus established: 1) the originator has privacy of the function; 2) each host has privacy in respect to their local input. The CCKM approach allows for autonomy in the agent path by creating an encrypted circuit that is a cascade of sub-circuits. Each host in the route of an agent's path would receive an encrypted circuit on which their input is applied. However, the CCKM protocol did not address the ability for each host to use the "unencrypted" local output of the agent because it was still encrypted and could only be evaluated by the originator.

Extending the CCKM approach further, Algesheimer et al. (2001) produced a non-interactive protocol (which we refer to as the ACCK protocol) similar to the trusted hardware of Loureiro and Molva (1999) that would allow for secure decryption of host output when CEF is used. The ACCK scheme, illustrated in Figure 1-a, makes use of a trusted generic computation service which is roughly equivalent to the trust we place in a public key infrastructure. To decrypt the output of the agent at the local host, the mappings for the semantics of the signals are encrypted with the public key of generic service. Each host accomplishes oblivious transfer with the generic service (instead of the originator who may be offline) to decrypt the signals for the output.

By using a secure middleman, the ACCK protocol allows inputs, outputs, and computations of all hosts to be hidden from the originator as well as any other host visited by the agent. The main assumption is that the trusted third party (TTP) does not collude with the originator or with any host, but as proposed would offer a generically secure service for any application.

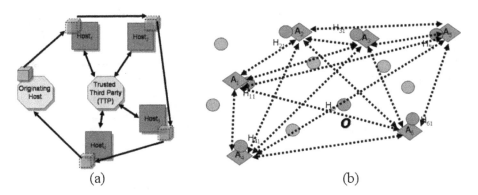

(a)   (b)

Fig. 1. Single and multi-round agent SMC

Two extensions have been proposed to ACCK that target replacement of the TTP: Zhong and Yang (2003) introduce a cryptographic primitive called verifiable distributed oblivious transfer (which we refer to as the VDOT protocol) and Tate and Xu (2003) introduce a multi-agent approach utilizing their oblivious threshold decryption (which we refer to as OTD). Figure 2-a shows a notional arrangement of parties in the VDOT scheme while Figure 2-b shows a notional arrangement of parties in the OTD approach.

In the VDOT protocol, mobile agent computations are divided into security-sensitive and non-security-sensitive portions. Code that requires integrity or confidentiality is transformed into a garbled Boolean circuit. Instead of interactions with one trusted third party, which has weaknesses involving the corruption of a single server to the detriment of the entire system, several trusted third party servers are used to replicate the functionality of TTP in the VDOT approach. VDOT guarantees with high probability the correctness of receiver's output, enforcement of the code and state privacy, protection from coalitions of malicious hosts and malicious TTPs, and the verification that servers give correct decryption of host signals.

Distribution of trust among a group of servers strengthens the original ACCK protocol and forces the table lookup for circuit signals to be performed by a group of servers that hold shares of the decryption. The VDOT protocol is general purpose in the sense that each host need only provide an interpreter for garbled circuits. By using distributed oblivious transfer, trusted third parties act as a proxy for agent owners and provide translation tables for host inputs without being able to discover host inputs themselves. Obvious disadvantages to the approach are increased communication complexity (which the authors contend is negligible in practice) and the complexity of breaking a program into security sensitive portions represented by a Boolean circuit.

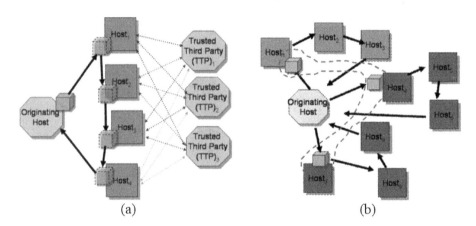

Fig. 2. Notional VDOT and ODT protocols

Tate and Xu's (2003) OTD protocol is similar in some regards to VDOT but actually eliminates the trusted third-party requirement altogether. As a primary distinction, their approach relies on multiple agents that are dispatched to disjoint sets of the possible host pool. Each of these agents act in a threshold manner (similar to VDOT) to decrypt the encrypted signals for a given host input without relying on the TTP. While the ACCK secure computation service overcame the interaction requirement of Yao's encrypted circuit evaluation–a limiting factor in the mobile code paradigm–OTD replaces this by means of cryptographic operations and multiple agents that cooperate together.

Multiple agents must agree before decryption of the host's input signals can occur and this in turn prevents cheating by keeping a list of hosts that have already decrypted a signal. Agents eventually return back to the originating host where all circuit results are decrypted and combined to produce a final result. The security in this method rests on the security of Yao's secure circuit evaluation, the security of the *1-out-of-2* oblivious transfer, and the strength of threshold cryptography. However, this protocol does not support free-roaming agents and requires knowledge of the set of hosts an agent will visit.

Though single-round non-interactive protocols reduce the communication overhead for SMC, message sizes increase proportionally, regardless of input or output size. Tate and Xu (2003), for example, state that it roughly takes 9k bytes to encrypt 32 bits of secret data under this scheme. Zhong and Yang (2003) mitigate overhead by keeping security sensitive portions separate from normal programmatic requirements. Using multiple round SMC offers another approach to accomplishing secure transactions with mobile agents, which we analyze now.

3.2 Multiple Round Approaches

Secure multiparty computations have a tradeoff between trust and efficiency. Neven et al. (2000) were one of the first to envision the use of mobile agents to *implement* SMC and reduce the overhead of the communication itself. Figure 3 summarizes four different approaches to integrating agents with hosts to accomplish SMC. Figure 3-a illustrates the ideal world where agents carry host inputs to a trusted third party and a protocol is evaluated without the expense of network broadcasts or bidirectional secure channels. In the context of the TTP, all parties can evaluate the protocol and the TTP is assumed to behave honestly with respect to host inputs.

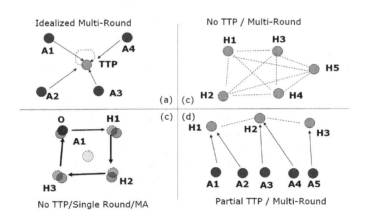

Fig. 3. Agent approaches to SMC

In the realm of mobile agents, as with many real world applications, it is preferable not to rely on a trusted third party and just perform an SMC among the parties of a function. Endsuleit and Mie (2003) utilize a group of multiple agents to support such an approach. Multiple agents carrying the same realized circuit are deployed to remote hosts where rounds of the secure protocols are evaluated among parties. Figure 1-b illustrates how such agents would be located on some set of hosts–working as a coalition to implement multi-agent computations based on an underlying SMC protocol. The extensive use of a broadcast channel is assumed here and the use of Ben-Or's protocol (1988) with an implementation of Shamir's (1979) secret sharing is suggested by Endsuleit and Mie as a possible SMC candidate. Endsuleit, later with Wagner(2004), suggest the use of more efficient protocols such as those of Hirt and Maurer's (2001).

The most secure but least efficient method is seen if Figure 3-b: here hosts simply become the execution environment and setup a multi-party protocol evaluation. In this case, both the computational and communicational complexity inherent in the chosen protocol must be faced and only high speed links (represented by the dotted lines) make such protocols practical. Single-round approaches discussed in the previous section are seen in Figure 3-c where an agent embodies the circuit to be securely evaluated and each host provides private input as the agent migrates. In Neven's approach (2000), a hybrid solution was posed as depicted in Figure 3-d where high speed communication links are present between one or more hosts. Participants in the n-party protocol send agents carrying their private inputs to one of these intermediate TTPs who can then efficiently and securely evaluate the function according to the rules of the protocol.

A nice feature of such multiple agent schemes is that any SMC protocol can be used as long as it meets the composable security properties defined by Canetti (2001). In order to adapt the Canetti model, which assume stationary parties, "slices" are defined as periods where a community of n agents is executed by a set of n different hosts with n migrations during that period. Resharing of data shares via the Ostravsky and Yung method (1991) is used to overcome the adverse affects of migration where malicious hosts can use acquired shares over time to compromise security.

The system supports self-repairing code and threshold agreement of computations, as long as up to $\frac{1}{3}$ of the community (agents or hosts) has not been compromised. SMC security results follow because Canetti establishes proof of a secure protocol for n parties computing a joint function in the presence of an active adversary corrupting up to some k limited servers. By using such agents to implement a redundantly shared global state of computation and coordinate activity, a wide variety of SMC protocols can be implemented. However, as with any multi-round solution, the communication complexity is extremely high and the originator must know a priori which hosts will be part of the computation.

4   Hybrid Approaches

SMC offers many advantages for securely accomplishing a group transaction. Several approaches by) define how an agent implements a circuit that is part of a multi-party computation. On one hand, Cachin et al. (2000), Algesheimer et al. (2001), and Zhong et al. (2003 define an originator that sends a single agent with a cascading circuit whose last migration signals the last computation of the circuit. Alternatively, Endsuleit with Mie(2003) and Wagner(2004) define an originator that sends multiple agents with the same circuit that executes protocols in stepwise multi-round fashion. Neven et al. (2000) uses a single or set of trusted execution sites to accomplish the SMC interaction. By combining these techniques where full protocols, multiple agents, and semi-trusted hosts are utilized, several advantages can be gained.

Malkhi et al. [2004] note a recent trend in SMC research where protocols are focused on specific application contexts–thereby allowing more efficient representations for specific tasks. This is true in the mobile agent paradigm as well where mobile agents will be used for specific tasks like auctions, trading, or secure voting. Feigenbaum et al. (2004) implement a secure computation mechanism utilizing SMC for collecting survey results with sensitive information. Their scheme uses data-splitting techniques and traditional Boolean circuit evaluation Yao-style (1986). Notably, it also uses a secure computation server, which acts in the role of a trusted entity within the system, and is the initiator of the 2-party function evaluation. We use this as an example to point out that in practical applications where true data privacy or true function privacy is needed, the presence of a trusted server is not beyond the realm of possibility. In fact, many agent applications which will be executed "in-house" will indeed benefit from the availability of such trusted entities.

Implementations of SMC in mobile agent systems must seek to reduce message size, number of broadcast or pairwise channels required, and the size of the circuit. To accommodate agent goals such as disconnected operations, the originator typically remains offline during the protocol evaluation. Agent autonomy requires the task to be accomplished by an agent that decides where and when to migrate. The requirement for full autonomy in the agent path and itinerary lends itself best to a combination of SMC that balances trust with efficiency. While there is a desire to eliminate the need or requirement for any trusted third party or trusted computation service (like PKI), some environments for SMC may be conducive to such assistance.

Non-interactive approaches are limited to a very small number of protocols that derive from Sander and Tschudin (1998) or Cachin et al (2000). Single-round approaches do not require trusted third parties but come with large message sizes and their own set of limitations which include reliance on a trusted entity similar to a PKI. Extensions to the non-interactive approach such as those by Tate and Xu (2003) and Zhong and Yang (2003) require foreknowledge of at least the number of hosts to be visited by the agent or the set of hosts themselves. As SMC protocols find better and more efficient means of expression over time (other than Boolean circuits), agent security approaches should be adaptable to integrate them as they improve. However, accomplishing multi-agent fully interactive protocols comes with stiff communicational costs.

We pose several hybrid approaches to SMC integration with mobile agents that can accommodate free-roaming itineraries as well as reduce overall communication cost. These approaches can be used to take advantage of the security properties of multi-party protocols while remaining flexible for integration of other protocols with higher efficiency in the future.

4.1   Invitation and Response

In our first approach, which we term "Invitation and Response", a multi-agent architecture is used with a form of semi-trusted execution sites. We define the protocol informally first and define two classes of agents: the invitation agent and the protocol agent. The originator, O, begins the task by sending an invitation agent which has some initial set of hosts to be visited or at a minimum the first host to be visited. Invitation agents are free-roaming and can make changes in their itinerary based on environmental conditions or information obtained from hosts or other information services.

To guard the invitation agent against data integrity and denial of service attacks, two different schemes can be used. First, a traditional data encapsulation technique can be used with the stipulation that the agent code itself is bound to the dynamic state of each agent instance. Data encapsulation protocols are reviewed by McDonald et al. (2005) and Jansen and Karygiannis (2000); Figure 4 depicts only one invitation agent being used. A second approach is to use multiple invitation agents with overlapping and redundant itineraries that reduce the possibility of malicious corruption. Each invitation agent has a uniquely identifiable code/state (to avoid replay attacks), but the collection of agents represents only a single uniquely identifiable task (such as a specific auction). If a host receives an agent requesting participation in the same unique event, it ignores subsequent requests much like network devices that only forward packets once.

Invitation agents carry with them the specifications for input corresponding to an originator's task. The specification represents the normal query for a host input which is part of a multiparty computation. Hosts will (or will not) respond to this invitation by dispatching a response agent. The response agent is based upon an underlying secure multi-party computation protocol and can be created in different ways.

First, the invitation agent can carry the code for the response agent which each host will use. The host will execute the response agent first on its local input and then send the response agent to a semi-trusted execution location to actually evaluate the circuit. The second approach involve the dynamic generation of the code and circuit by the invitation agent when a host responds positively. A third method would involve each host responding to the invitation by sending its input encrypted to the semi-trusted execution site. This method resembles the traditional notion of the ideal SMC environment where parties send their input to a TTP for execution of the protocol.

Regardless of the method chosen, response agents migrate and move to a set of semi-trusted host environments in order to evaluate the protocol. The semi-trusted hosts can be specifically designed to serve multi-party computations (predefined based on some underlying protocol) or can simply provide basic agent execution environments with communication facilities. The key characteristic

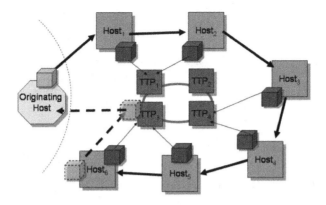

Fig. 4. Invitation and response protocol

of these hosts are that they are connected by a high bandwidth network so communication costs are negligible. This corresponds to the SMC approach seen in Figure 3-d where a tradeoff is made with overhead by bringing agents closer together through the availability of a high speed communication link among the servers. Environments are semi-trusted because group and threshold operations can be accomplished to eliminate the full trust in any one server.

In terms of security, "invitation and response" has the following properties. Hosts can only send one agent to the computation which removes the possibility the circuit can be evaluated on multiple host inputs. As long as multiple host submissions (and therefore cheating) are detectable, the originator's privacy is preserved. The local host input is kept private under two scenarios: 1) when the execution sites are fully trusted, as depicted in Figure 5-a, no extra security is required and each execution site is expected to maintain privacy of host inputs; 2) when the execution sites are semi-trusted, as depicted in Figure 5 a threshold mechanism can be used to distribute the trust among the set of hosts for decryption operations of circuit operations.

The advantages of this hybrid approach include the ability to accommodate true free-roaming agent scenarios and to use any type of secure multi-party protocol for the evaluation of the secure function. Protocols which have high communication and low computational complexity can thus be favored because agents are sent to a semi-trusted environment that has an assumed high speed link among execution sites. Depending on the trust level of the application environment, fully trusted hosts may be a possibility and simpler protocols can be utilized that do not involve threshold decryption of signals.

The selection of execution environments becomes one of the issues with the invitation and response protocol. The two primary factors are the presence of a high speed communications link between servers and a common trust level among all parties of the protocol with the trusted servers. Migration of agents also becomes more structured as the only free-roaming portion of the task is to find interested parties to the computation itself. Response agents only make two

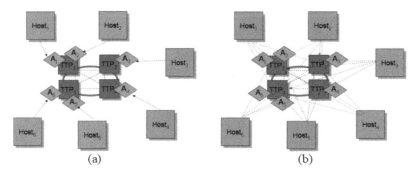

Fig. 5. Fully-trusted (a) and semi-trusted (b) evaluation

subsequent migrations: to the trusted server environment and then back to the originator, who can decrypt the final agent state and obtain the result.

One of the issues well discussed by Algesheimer (2001) with SMC and agents is how can an individual host get its local output as the mobile agent migrates. In invitation and response, local host output can be handled in one of two ways. First, since the host output is not private in terms of the originator, O can be responsible for providing the output to each host after the evaluation of the secure function on the execution environment and after response agents migrate back to the originator. Second, the set of TTPs can each send their share of the output or the single TTP can send the output corresponding to a host back to it, through message passing or another class of agent.

4.2   Multi-agent Trusted Execution

When the itinerary of an agent is *known* beforehand, simpler agent architecture can be used to facilitate execution. Several configurations are possible for host environment in terms of a secure computation. First the host can be the computation environment for a cascaded circuit that requires only one round of execution. Next, the host can communicate with a semi-trusted party to evaluate an encrypted circuit or can communicate with a *threshold* of semi-trusted parties that provide signal decryption services in an oblivious manner. The host can also be the computation environment for a multi-round circuit and can be visited by more than one agent.

Multiple agents can be used to initiate a multi-party protocol among a predefined set of hosts. Similar to the multi-agent approaches of Endsuleit (2003,2004), multi-agent trusted execution would allow agents to migrate to hosts where input is first gathered. When one trusted execution environment is fully trusted by all parties, agents can then migrate there to accomplish a multi-round protocol, as suggested by Neven et al. (2000).

If a less trusted set of execution environments are in view, trusted servers need to be linked by a high bandwidth communications network. In either case, agents are dispatched to hosts in the manner of a traditional multi-party function as the first step of the task. Once agents obtain host input they then migrate to a centralized trusted execution site where the multi-round protocol is evaluated. In performing such an operation, the goal is to minimize the communication overhead of the network while maximizing the benefit of *any* given SMC protocol chosen.

5 Conclusions

There is a distinct tradeoff when using secure multiparty computations with mobile agent applications. The overhead of both computation and communication are barriers which must be overcome before protocols can be used in a practical manner. We have reviewed the state of the art in such integration approaches and posed variations of hybrid approaches that utilize fully trusted or semi-trusted execution environments for secure multi-agent computations. These schemes offer an alternative to other architectures posed which combine the best of non-interactive approaches and multi-round SMC approaches. Future work will involve the formal description of such protocols and an analysis of their overhead when specific SMC protocols are in view.

Acknowledgements

I would like to thank Dr. Mike Burmester for his contributions and insightful discussions regarding multi-party computations.

References

Abadi, M., Feigenbaum, J.: Secure Circuit Evaluation: A Protocol Based on Hiding Information from an Oracle. J. of Crypto. **2** (1990) 1–12

Abadi, M., Feigenbaum, J., Kilian, J.: On hiding information from an oracle. J. of Comp. and Sys. Sci. **39**(1989) 21–50

Algesheimer, J., Cachin, C., Camenisch, J., Karjoth, G.: Cryptographic security for mobile code. In: Proc. of the 2001 IEEE Symp. on Security and Privacy. (2001) 2–11

Bellare, M., Micali, S.: Non-Interactive oblivious transfer and applications. Advances in Cryptology-CRYPTO '89 (1990) 547–559

Bellare, M., Micali, S., Rogaway, P.: The round complexity of secure protocols. In: Proc. of 22nd Annual ACM Sym. on Theory of Comp. (STOC) (1990) 503–513

Ben-Or, M., Canetti, R., Goldreich, O.: Asynchronous secure communications. In: Proc. of 25th Annual ACM Symp. on Theory of Comp. (1993) 52–61

Ben-Or, M., Goldwasser, S., Wigderson, A.: Completeness theorems for non-cryptographic fault-tolerant distributed computation. In: Proc. of Annual ACM Symp. on Theory of Comp. (1988) 1–10

Ben-Or, M., Kelmer, B., Rabin, T.: Asynchronous secure computations with optimal resilience. In: Proc. of 13th Annual ACM Symp. on Principles of Distributed Computing (PODC). (1994) 183–192

Bierman, E., Cloete, E.: Classification of malicious host threats in mobile agent computing. In: Proc. of the 2002 Annual Research Conference of the South Africa IoCS and IToETT. Port Elizabeth, South Africa (2002) 141–148

Cachin, C., Camenisch, J., Kilian, J., Mller, J.: One-round secure computation and secure autonomous mobile agents. In: Proc. 27th Int'l Colloquium on Automata Languages and Programming, LNCS **1853**, Springer-Verlag (2000) 512–523

Canetti, R.: Universally composable security: A new paradigm for cryptographic protocols. In: Proc. of the 42nd IEEE Symp. on Foundations of Comp. Sci. (2001) 136

Canetti, R.: Security and composition of multiparty cryptographic Protocols. J. of Cryp. **13:1** (2000) 143–202

Chaum, D., Crpeau, C., Damgard, I.: Multiparty unconditionally secure protocols (Extended Abstract). In: Proc. of the 20th Annual ACM Symp. on Theory of Comp. (1988) 11–19

Cramer, R., Damgard, I., Dziembowski, S., Hirt, M., Rabin, T.: Efficient multiparty computations with dishonest minority. In: Proc. of EUROCRYPT '99, LNCS **1592**, Springer-Verlag (1999) 311–326

Cramer, R., Damgard, I., Nielsen, J.B.: Multiparty computation from threshold homomorphic encryption. In: Advances in Cryptology - EUROCRYPT '01, LNCS **2045**, Springer-Verlag (2001) 280–300

Chaum, D., Damgard, I., Van De Graaf, J.: Multiparty computations Ensuring privacy of each party's input and correctness of the result. In Pomerance, C. (ed.): Proc. of CRYPTO '87, LNCS **293**, Springer-Verlag (1988) 87–119

Damgard, I., Nielsen, J.: Universally Composable Efficient Multiparty Computation from Threshold Homomorphic Encryption. In: Advances in Cryptology - Crypto 2003, LNCS **2729**, Springer-Verlag (2003) 247–264

Endsuleit, R., Mie, T.: Secure multi-agent computations. In: Proc. of Int'l Conf. on Security and Mngmnt. (2003) 149–155

Endsuleit, R., Wagner, A.: Possible attacks on and countermeasures for secure multi-agent computation. In: Proc. of Int'l Conf. on Security and Management (2004) 221–227

Feigenbaum, J., Pinkas, B., Ryger, R., Saint Jean, F.: Secure computation of surveys. In: EU Workshop on Secure Multiparty Protocols (2004)

Fitzi, M., Garay, J., Maurer, U., Ostravsky, R.: Minimal Complete Primitives for secure multi-party computation. J. of Crypto. **18** (2005) 37–61

Goldreich, O.: Secure multi-party computation. Working Draft, Version 1.2. (2000)

Goldreich, O., Micali, S., Wigderson, A.: How to Play Any Mental Game. In: Proc. of the 19th Annual ACM Symp. on Theory of Comp. (1987) 218–229

Hirt, M., Maurer, U.: Robustness for free in unconditional multi-party computation. In: Proc. of Cypto'01, LNCS **2139**, Springer-Verlag, (2001) 101–118

Jansen, W., Karygiannis, T.: NIST Special Publication 800-19 - Mobile sgent decurity. National Institute of Standards and Technology (2000)

Kilian, J.: Founding cryptography on oblivious transfer. In: Proc. of 20th Annual ACM Symp. on Theory of Computing (1988) 20–31

Loureiro, S., Molva, R.: Function hiding based on error correcting Codes. In: Proc. of the 1999 Int'l Wrkshp on Cryptographic Techniques and E-Commerce, CrypTEC '99, City University of Hong Kong Press (1999) 92–98

Malkhi, D., Nisan, D., Pinkas, B., Sella, Y.: FAIRPLAY-a secure two-party computation system. In: Proc. of 2004 USENIX Security Symp. (2004) 287–302

McDonald, J. T., Yasinsac, A., Thompson, W.: Taxonomy of mobile Agent Security, submitted ACM Computing Surveys (2005). Available, http://www.cs.fsu.edu/research/reports/TR-050329.pdf

Micali, S., Rogaway, P.: Secure computation. In: Advances in Cryptology-CRYPTO '91, LNCS **576**, Springer-Verlag (1992) 392-404

Naor, M., Nisim, K.: Communication complexity and secure function evaluation. In: Electronic Colloquium on Computational Complextiy (ECCC) **8** (2001) 62

Naor, M., Pinkas, B.: Efficient oblivious transfer protocols. In: Proc. of SIAM Symposium on Discrete Algorithms, Washington, D.C. (2001) 448–457

Naor, M., Pinkas, B.: Distributed oblivious transfer. In: Advances in Cryptology – Asicrypt '00, LNCS **1976**, Springer-Verlag (2000) 200–219

Naor, M., Pinkas, B., Sumner, R.: Privacy preserving auctions and mechanism design. In: Proc. of 1st ACM Conf. on Electronic Commerce (1999) 129–139

Neven, G., Van Hoeymissen, E., De Decker, B., Piessens, F.: Enabling secure distributed computations: Semi-trusted hosts and mobile agents. Net. and Info. Sys. J. **3:43** (2000) 1–18

Ostravsky, R., Yung, M.: How to withstand mobile virus attacks. In: Proc. of 10th Annual ACM Symp. on Principles of Distr. Comp. (1991) 51–59

Rabin, T., Ben-Or, M.: Verifiable secret sharing and multiparty protocols with honest majority. In: Proc. of the 21st Annual ACM Symposium on Theory of Comp., Seattle, Washington, USA (1989) 73–85

Rivest, R. L., Adleman, L., Dertouzos, M. L.: On data banks and privacy momomorphisms. In Demillo, R. A. et al. (eds.): Foundations of Secure Computation, Academic Press (1978) 169–177

Sander, T., Tschudin, C. F.: Protecting mobile agents against malicious hosts. In: Mobile Agent Security, LNCS **1648**, Springer-Verlag (1998) 44–60

Shamir, A.: How to share a secret. Comm. of the ACM **22:11** (1979) 612–613

Tate, S. R., Xu, K.: Mobile agent security through multi-agent cryptographic protocols. In: Proc. of the 4th Int'l Conf. on Internet Comp. (2003) 462–468

Yao, A. C.: How to generate and exchange secrets. In: Proc. of the 27th IEEE Symposium on Found. of Comp. Sci. (1986) 162–167

Yokoo, M., Suzuki, K.: Secure multi-agent dynamic programming based on homomorphic encryption and its application to combinatorial auctions. In: Proc. of the 1st Int'l Joint Conf. on Autonomous Agents and Multiagent Systems, Bologna, Italy (2002) 112–119

Zhong, S., Yang, Y. R.: Verifiable distributed oblivious transfer and mobile agent security. In: Proc. of the 2003 Joint Workshop on Foundations of Mobile Comp., ACM Press (2003) 12–21

An XML Standards Based Authorization Framework for Mobile Agents

G. Navarro and J. Borrell

Dept. of Information and Communications Engineering
Universitat Autònoma de Barcelona, 08193 Bellaterra, Spain
{gnavarro, jborrell}@ccd.uab.es

Abstract. An outstanding security problem in mobile agent systems is resource access control, or authorization in its broader sense. In this paper we present an authorization framework for mobile agents. The system takes as a base distributed RBAC policies allowing the discretionary delegation of authorizations. A solution is provided to assign authorizations to mobile agents in a safe manner. Mobile agents do not need to carry sensitive information such as private keys nor they have to perform sensitive cryptographic operations. The proposed framework makes extensive use of security standards, introducing XACML and SAML in mobile agent system. These are widely accepted standards currently used in Web Services and Grid.

Keywords: Mobile Agents, Authorization, Access Control, XACML, SAML.

1 Introduction

During the last years, mobile agent technologies have witnessed an steady, if not fast, increase in popularity. Probably, the main hurdle to a wider adoption are the security issues that mobility brings to the picture [7]. Among them, an outstanding one is resource access control. Traditional access control methods rely on the use of centralized solutions based on the authentication of global identities (for example, via X.509 certificates). These methods allow to explicitly limit access to a given resource through attribute certificates or Access Control Lists, and rely on a centralized control via a single authority. Despite providing effective means of protection, these techniques suffer from serious drawbacks; in particular, they give raise to closed and hardly scalable systems. Practical mobile agent systems demand lightweight, flexible and scalable solutions for access control, in order to cope with the highly heterogeneous nature of their clients. In the same vein, solutions depending on centralized entities (such as traditional Certification Authorities) should be avoided.

Recent developments in the area of access control, in an attempt to further ease access control management, have brought into the picture Role-based Access Control (RBAC) [11]. In these schemes, privileges of principals requesting access to a resource are determined by their membership to predefined roles. The use of RBAC greatly simplifies access control management and is specially suited to mobile agents scenarios, where agents privileges are subsumed in a possibly more general RBAC system.

M. Burmester and A. Yasinsac (Eds.): MADNES 2005, LNCS 4074, pp. 54–66, 2006.

Even in RBAC environments, there may be some situations where more flexibility is required. Discretionary delegation of authorizations between users may provide flexibility beyond RBAC policies. A user may temporary delegate some rights to another one without having to modify the system policies. Delegation of authorizations has been successfully introduced by *trust management* systems such as *Simple Public Key Infrastructure/Simple Distributed Secure Infrastructure* (SPKI/SDSI) [9] and KeyNote [3].

This article presents an access control framework for mobile agents. We combine RBAC and discretionary delegation of authorizations to provide a flexible, and distributed system for access control in such scenarios. In our approach, mobile agents do not carry explicit information regarding resources access control, avoiding the privacy concerns associated with sensitive data embedded in mobile code. We have implemented our proposed scheme on MARISM-A, a JADE-based agent platform. Our framework is called XMAS (*XML-based Mobile agents Authorization System*).

In Section 2 we explain the motivations behind XMAS. Sections 3 and 4 describe the main naming and authorization issues. We describe the main components and functionality of XMAS in Section 5. Section 6 details how roles are assigned to mobile agents. Finally, Section 7 summarizes our conclusions.

2 Motivations and Related Work

Proposals for access control in multiagent systems supporting agent mobility are scanty. Most of the security work in mobile agents deals with protecting communications, itineraries, and so on, but few of them deal with the protection of resources and the management of access control rights.

Usually, proposed systems rely on ad-hoc and centralized solutions. As an example, in [23] the authors propose a solution based on proprietary credentials, which are carried by the agent as authorizations. These credentials are combined with the Java protection domains to provide resource access control. The solution is interesting but it relies too much on the platform itself, and more precisely in Java. The system cannot be applied to other platforms and it is not distributed, since it requires a centralized policy. A similar approach was adopted in [21]. And most notably [14].

JADE (http://jade.tilab.it) also provides a security add-on[13], which is also based on the Java security architecture and does not support agent mobility. FIPA (Foundation for Intelligent Physical Agents: http://fipa.org) also began to consider agent security through a technical committee, but the work has not been finished, and there is no FIPA recommendation at the moment on security related topics.

Other mobile agent platforms, such as NOMADS [21], provide access control based on Java-specific solutions and eventually used KAoS to provide high level policy languages and tools. KAoS [4] is a high level policy language currently focused on semantic web services. Despite its complexity, KAoS does not support mobile agents.

On the other hand, several communities have substantially contributed to access control and authorization management in distributed environments, most notably in Web Services and Grid. Systems such as Shibboleth [10], PRIMA [16], PERMIS [6], or Cardea [15], are just an example. These systems tend to be policy-based and provide support for Role-based Access Control (RBAC).

Another important initiative has been the development of standards for expressing authorization information. For instances the *eXtensible Access Control Markup Language* (XACML)[22] is a general purpose access control policy language specified in XML. XACML is intended to accommodate a wide variety of applications and environments. Together with the policy specification language it also provides a query and response format for authorization decision requests. The expressiveness, semantics for determining policy applicability, support for advanced features, and the fact that it was designed with distributed environments in mind, makes XACML a great standardized base for large-scale authorization frameworks.

At the same time, the *Secure Assertion Markup Language* (SAML)[19], provides a standard XML-based framework for exchanging security information between online business partners. Security information is exchanged in form of *assertions*. SAML provides three types of *assertion statements*: Authentication, Attribute, and AuthorizationDecision. Broadly speaking an assertion has an *issuer*, a *subject* or *subjects*, some *conditions* that express the validity specification of the assertion, and the *statement*. The assertion may be signed by the issuer. SAML also provides query/response protocols to exchange assertions and bindings over SOAP and HTTP.

XACML and SAML are being widely adopted by the previous projects in web and grid services, as well as other industry solutions (Liberty Alliance, for instance). These two standards are also well suited to coexists together. For instance, Cardea uses SAML protocols in an XACML policy based environment. In XMAS, we have choose to adopt a combination of XACML and SAML. XACML is used to express the main authorization policy, which is an RBAC policy. While SAML is used to express discretionary delegation statements and provides the protocols to exchange all the information.

This work is the continuation of a previous framework for authorization management in mobile agent systems [17], based on SPKI/SDSI. From this previous experience, we have developed a more scalable system making use of current standards and introducing a more powerful delegation model, among other things.

The main contributions of XMAS is to provide a novel authorization framework for mobile agents. It presents the combination of RBAC with discretionary delegation of permissions or authorizations. The proposed use of XACML and SAML standards in mobile agent system, will allow agents to interact with other systems such as Web Services or Grids. We also describe a safe way to assign roles to mobile agents, so the mobile agent does not need to carry sensitive information.

3   Naming Schema

XMAS uses a distributed naming schema strongly influenced by SPKI/SDSI, which is used to create and manage roles and groups. Each entity, (agent, platform, human user, etc.) is denoted as a *principal*. All principals (including static agents) have a pair of cryptographic keys, except mobile agents. The public key acts as a global identifier of the principal. In order to make it more manageable one can use the *hash* of the public key as an abbreviation for the public key. Each principal generates by itself the keys and is responsible of its own identity, so there is no need for a centralized CA, although it can be used if it is needed. Mobile agents are identified by a hash of its code.

A principal can define local names of entities under its own responsibility. In order to do that, an entity has an associated local name space called *name container*. The name container has entries of the type: (<principal>,<local-name>). The *principal* corresponds to the principal for whom the local name is being defined, and *local-name* is an arbitrary string. The *principal* can be specified as a public key or as a *fully qualified name* (see below).

For example, consider a principal with public key PK_0, which creates an agent with public key PK_1 and wants to name it *my-agent*. The name container of the entity will have an entry of the form: (PK_1, my-agent). Now on, the agent PK_1 can be referenced by the name *my-agent* in the local name space of PK_0. An important issues is that a third party can make a reference to a name defined in other name containers through a *fully qualified name*. A name container is identified by the public key of the owner, so the fully qualified name PK_0 *my-agent* makes reference to the name *my-agent* defined in the name container of PK_0. It is important to note that given the properties of cryptographic keys, it is commonly assumed the uniqueness of the public key, so fully qualified names are globally unique.

These names, make it very easy for a principal to create groups or roles. For instances, a user PK_{admin} can create a group *employees* with members PK_a, PK_b and the agent PK_1, by adding the following entries in its name container:

$(PK_a,\ employee)$

$(PK_b,\ employee)$

$(PK_0\ my\text{-}agent,\ employee)$

In our framework names are expressed not only as identifiers of a principal but also as *attributes* in the case of roles. A name container entry can be expressed as a SAML assertion, where the issuer is the owner of the name container, the subject is the principal and the name is expressed as an AttributeStatement. We denote such an assertion as:

$$\{(PK_{admin}\ security\text{-}staff,\ employee)\}_{PK_{admin}^{-1}}$$

where PK_{admin}^{-1} denotes the private key corresponding to the public key PK_{admin}, which digitally sings the assertion determining the issuer or the owner of the name container where the name is defined (note that this assertion can only be issued by the owner of the container). The assertion may also contain validity conditions, which are not shown for clarity reasons.

Although any principal may create and manage roles for its own purpose, in order to make them available to the XMAS framework, they need to be introduced into an XACML policy managed by a special entity, the Role Manager (see Section 5.2).

4   Delegation of Authorization

Authorizations may be assigned to principals either through an XACML policy rule or through a SAML assertion. An authorization will have an issuer, the principal granting the authorization, which in the case of an XACML rule will be the policy owner, and in the case of a SAML assertion will be the issuer of the assertion. It also has a subject, and the authorization itself.

A key point of XMAS is the ability to allow *delegation* of authorizations between principals. Furthermore, we adopt a *constrained delegation model*, which provides a powerful delegation mechanism. It allows to differentiate between the right to delegate an authorization and the authorization itself. That is, a principal can have the authority to delegate a given authorization but may not be able to hold (and use) the authorization. This is different from the approach in *trust management* systems [3,9] in which the right to delegate a privilege can be given only to those that have the privilege for themselves. A formal definition and a full description of the constrained delegation model can be found in [2,20].

In short, there are two types of authorizations:

- *Access-level authorization*: defines an access-level permission such as *read file*, assigned to a principal. It is denoted as: $authz(s, a)$, where s is the subject (a principal) and a is the specific access-level permission.
- *Management-level authorization*: defines the authority to declare an access-level authorization. That is, the right to delegate an access-level authorization. We denote it as: $pow(s, \phi)$, where s is the subject (a principal) and ϕ is either an access-level authorization or another management-level authorization.

A complete authorization can be denoted as:

$$\{(K_s,\ p)\}_{K_i^{-1}}$$

Where the subject K_s receives the authorization p from the issuer K_i. Again, K_i^{-1} denotes the private key associated with K_i, which issues the authorization. Both names and authorizations are denoted with a 2-tuple, the main difference between them being the second element. In the case of an authorization it will be of the form of: $auth(\ldots)$, or $pow(\ldots)$.

The flexibility and the power introduced by this constrained delegation model comes with a cost. In order to determine if a principal has gained an authorization through delegation, we have to find a delegation chain from a source of authority to the principal. Finding this *authorization proof* may become very complex. In order to simplify it we rely on a combination a *privilege calculus* [20], to find authorization proofs in the presence of *constrained delegation*, and the name resolution and reduction rules of the SPKI/SDSI certificate chain discovery algorithm[8]. Broadly speaking, in order to find delegation chains, we use the name resolution algorithms to resolve all the names into public keys (or hashes in the case of mobile agents), and then apply the privilege calculus to find the authorization proof. The procedure is computationally feasible presenting a polynomial order of complexity, an it is performed by an special entity, the delegation Assertion Repository Manager (see Section 5.4).

5   XMAS Components

The XMAS system is implemented on top of MARISM-A [18], a secure mobile agent platform implemented in Java. It provides extensions on top of the JADE system. JADE implements the standard elements of the FIPA specification and provides additional services for the management of the platform. Mobility is achieved by the MARISM-A mobility component which is integrated into JADE. On top of JADE there are the main MARISM-A components such as the authorization framework presented in this paper, and other MARISM-A services such as: cryptographic service, directory service, service discovery, etc.

Agents in MARISM-A can be mobile or static, depending on the need of the agent to visit other agencies to fulfill its task. There are several types of mobile agents according to the characteristics of its architecture: basic or recursive structure, plain or encrypted, itinerary representation method, etc. Agents can communicate with each other through the agency communication service.

XMAS is made up of five independent components, which interact to perform all the required functionality. This components are implemented as static agents. The messages exchanged between these components are SAML protocol messages, enclosed in FIPA Agent Communication Language, and using FIPA ontologies. The SAML protocols already provide bindings for transport over SOAP and HTTP, which are used to communicate XMAS components with entities outside the multi-agent domain, such as external authorities, normally in the form of web services or grid services, where the use of SAML and SOAP is widespread. The reason why we choose FIPA ACL is that it has become an standard in multi-agent environments. Most of the existing agents platforms support the FIPA specifications, which provide transport over IIOP, HTTP, and WAP.

In order to locate required information and modules. XMAS relies on a service discovery infrastructure. It is provided by the underlying agent-platform implementing the FIPA agent discovery specification [12]. This is specially relevant in the presence of mobile agents because an authorization decision may involve the gathering of information from components located in different platforms. In the case of XACML, the XACML policies already provide mechanisms to easily distribute policies through references, without the need for discovery services. But for example, the delegation chain discovery can use the service to locate an specific authorization manager.

5.1   Authorization Manager (AM)

The *Authorization Manager* (AM) manages the assignment of a set of specific authorizations to specific roles. It may also provide the ability to delegate authorizations to other AMs in order to distribute the authorization management. Since the authorization policy is local to the AM agent, it does not need to follow any specification and its format could be implementation-dependent. Even so, we use XACML policies to provide a standardized, and uniform approach.

The AM has two different local policies, expressed in XACML. The first one is the *XML Authorization Policy* (XAP). The XAP specifies the authorization assignment to roles. In other words determine, which roles hold which permissions or authorizations. The second one is the *Administrative XML Authorization Policy* (XAP-*adm*). The XAP-*adm* determines the administrative authorities for the authorizations managed by the AM. An AM may receive a request to issue an authorization for an specific role, if the authorization comes (directly or indirectly) from one of the authorities for that specific authorization, it is granted and the XAP is modified accordingly.

Principals sending requests to the AM will normally be other agents using SAML protocols. Even so, we have provided a GUI tool for the easy specification of the AM policies by human administrators: the *AM-console*.

As Figure 1 shows, the XAP is possibly the most complex policy of the system. Originally we used a proprietary definition of the XAP policy, but we have recently adopted the proposal on RBAC profile for XACML [1]. It is currently a Committee Draft in the XACML version 2.0 specification.

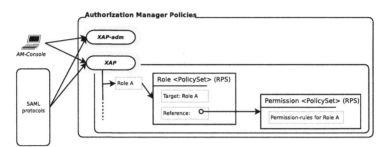

Fig. 1. Authorization Manager (AM) policies

The XAP is made up of several <*PolicySet*>s. For each role there is a *Role* <*Policy-Set*> (RPS) associating the role to a *Permission* <*PolicySet*> (PPS), which determines the actual permissions associated with the role. It is important to note that the entry point to query the XAP is the RPS and never the PPS. Or in other word, a PPS cannot be directly consulted, the only way to access it is through an RPS. This allows to support role hierarchies, ensuring that only principals of the given role can gain access to the permissions in the given PPS. Role hierarchy is thus defined by including a reference to the PPS associated with the junior (or sub) role in the PPS of the senior (or super) role.

5.2 Role Manager (RM)

The *Role Manager* (RM) manages a set of roles, mainly role membership (Section 6 details how roles are assigned to mobile agents).It has an *XML Role Policy* (XRP) and an *Administrative XML Role Policy* (XRP-adm). The XRP-*adm* policy specifies administrative role authorities. That is, principals that can request the creation of roles, the assignment of principals to roles, or the specification of constraints on the role assignment.

The XRP policy has two types of XACML policies. It has policies to define role membership and a *Role Assignment Policy*. The role membership is expressed as an XACML Attribute, associated to the given principal (see Section 3). On the other hand, the Role Assignment Policy, is used to answer the question "Is subject X allowed to have role R_i enabled?"[1].

As in the case of the AM, there is a GUI tool (*RM-console*) specially designed so RM policies can be managed by a human, although they are mainly intended to be managed by autonomous agents through SAML protocols.

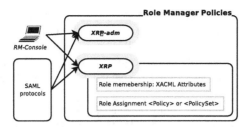

Fig. 2. Role Manager (RM) policies

5.3 Resource Controller (RC)

The *Resource Controller* (RC) main task is to control the access to a set of resource. It acts as a Policy Enforcement Point [24], receives access requests from principals and allows or denies the request depending on authorization responses from the Authorization Decision Engine.

5.4 Delegation Assertion Repository Manager (ARM)

In XMAS, principals may delegate authorizations in a discretionary way as SAML assertions. This assertions need to be gathered to perform the authorization decision. Although each principal can maintain a local cache of issued assertions, having to search and query all principals for assertions relating a given request can be too expensive in terms of communications due to the high complexity and possibilities imposed by possible delegation paths.

The *delegation Assertion Repository Manager* (ARM), keeps and updated repository of SAML delegation assertions. When a principal issues a delegation assertion it stores the assertion in the repository. Furthermore the ARM can find partial *authorization proofs* by finding delegation chains from the repository (see Section 4). This way we solve the problems derived from assertion distribution and leave the task to perform chain discoveries to the ARM and not to the other principals. It decreases communication traffic, assertions do not need to travel constantly from one principal to another, and reduces the task that generic principals need to perform.

5.5 Authorization Decision Engine (ADE)

The *Authorization Decision Engine* (ADE) is the main responsible for determining authorization decisions acting as a Policy Decision Point [24]. Given an authorization request, normally coming from an RC, the ADE is able to determine if the request has to be granted or denied by retrieving information from required AMs, RMs, ARMs, and other possible sources of information such as external Attribute Authorities.

An important issue with the ADE is that it is unique for a single agent platform. XMAS allows to place any number of AM, RM, ARM and RC in a single platform, they may be set up by different principals, or for different applications. But there is only one ADE in the platform. This ADE is the one used by all the RCs in the platform. Note that there is an implicit trust relation between the resources or service owners and the platform where they are placed.

Conflict Resolution. Delegation of authorizations was introduced to provide more flexibility in the system, but at the same time it may introduce possible inconsistencies. It is possible to find conflicts between the RBAC policies and the delegation assertions. For instance, we can find a principal who has received an authorization from a delegation assertion, but at the same time has enabled a role, which explicitly denies such authorization.

We adopt a *closed policy*, all authorizations are denied by default, and the policies express permitted authorizations. In XACML it is possible to specify deny rules, but this rules are normally used to express exceptions and are not used in a normal base. Even so, the ADE provides a conflict resolution procedure to resolve possible inconsistencies. The ADE can obtain three different evaluation results: an XACML deny or XACML permit from the RBAC policies, and an explicit permit from delegation assertions. The ADE uses the following order of precedence:

begin
    if *(XACML-decision == Deny)* **then** return Deny;
    if *(XACML-decision == Permit)* **then** return Permit;
    if *(there is a delegation chain == Permit)* **then** return Permit;
    else return Deny;
end

Algorithm 1. ADE combining algorithm

Note that negative rules from the RBAC policies supersede the delegation statements. It is possible to apply constraints on delegations in the RBAC policies. For example it is possible to define a deny rule in the RBAC policies for a role, so principals cannot delegate authorizations received as a result of being members of that role.

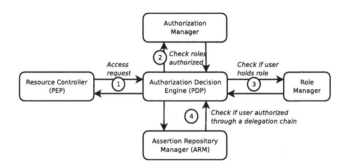

Fig. 3. ADE data flow

Figure 3 shows a simplified data flow for the decision engine. Note that, for example, querying the AM to check the roles authorized, may involve the interaction with several AMs depending on the one that handles the specific authorization.

6   Establishing Mobile Agents Role Membership

One of the first problems we found when planning the authorization model, is if a mobile agent should have a cryptographic key pair and be considered as a principal. A mobile agent cannot trivially store a private key, as well as perform cryptographic operations such as digital signatures. There are some proposals to store sensitive information (private keys) in mobile agents [5]. But the problem arises when the mobile agent uses the private key to compute a cryptographic operation. The agency where the agent is in execution will be able to see the private key or at least, reproduce the operation. As a result we consider that a mobile agent should not have a private key.

Since mobile agents cannot have private keys, we can not delegate authorizations to a mobile agent or make it member of a role. Our approach is to set, as member of the role, a hash of the agent's code. A principal can be identified by a a public key or a hash of a public key. So a hash may be seen as a principal, subject of a certificate. This is even supported by certificate frameworks such as SPKI/SDSI or even X.509, where a hash of an object can be considered as a principal.

In order to establish the role membership of a mobile agent we consider two different approaches. To show them we use a simple example where a user K_u is member of the role *physician* defined by a Role Manager RM_0 of the agency i, which has a resource controlled by a given RC. The role *physician* allows its members to access the given resource in agency i. The user has a mobile agent with the code m_i to be executed in platform i. The goal is to set the hash of m_i as member of the role *physician*.

User-managed Role. RM_0 makes member of the role *physician* a role (or group) defined by the user K_u, say *agent*. Then, the user K_u can make member of its role *agent* any hash of agent's code:

$$\{(K_u \ agent, \ physician)\}_{RM_0^{-1}}$$
$$\{(hash(m_i), \ agent)\}_{K_u^{-1}}$$

RM-managed Role. RM_0 makes member of the role *physician* the user K_u. Then the users sends a request to the RM_0 to set the agent code's hash as member of the role:

$$\{(K_u, \ physician)\}_{RM_0^{-1}}$$
$$\{(hash(m_i), \ physician)\}_{RM_0^{-1}}$$

The *user-managed role*, is quite straightforward, gives the user full flexibility to manage the role K_u *agent*, and does not require any special mechanism or protocol. The user may add to the role any agent (or even user) she wants. The main problem with this first approach is related to the accountability of the system. In e-commerce application there will be different degrees of trust between users. For example, a hospital may trust an internal physician to manage its role *agent*, but a client from an external research center may not be so trusted. In the first case we will use a *user-managed role*, while in the second one we will use an *RM-managed role*.

This last approach requires an additional protocol, which is implemented as a SAML protocol. The user sends an agent role assignment request to the RM, including m_i. The RM verifies the client's request, if permitted, it computes the hash of m_i and issues the corresponding XACML attribute, in its XRP.

The RM may store the code m_i to use it for further security audits (or a hash of the code). Note that this approach does not allow the user to manage the role and extend it to other agents. The user needs to send a request for each agent.

When the mobile agent arrives to agency i, it will send an access request to the RC controlling the resource. The RC just has to compute the hash of the code m_i and check, through a request to the ADE, if the agent is authorized to access the resource.

The main drawback of this approach is that a mobile agent is not capable of issuing a delegation assertions by itself, since the agent can not sign them. But note that this does not mean that the agent cannot issue or delegate an authorization, which may be certified by a trustee.

Authorizations associated to an agent are normally determined by its role membership. This way we can say that the agent will have *dynamically assigned authorizations* during its lifetime. If the authorizations associated with a role change, the authorizations related to the agent also change.

7   Conclusions

In this paper we have presented a novel authorization framework for mobile agent systems, XMAS. The main contributions in the field of mobile agents can be summarized in three points. First, it presents an RBAC-like distributed policy and at the same time adds support for discretionary delegation of authorizations. Second, the framework makes extensive use of current security standards for distributed systems, more specially XACML and SAML, which easies the interoperation with other systems such as Web Services or Grid. Finally we propose a mechanism to authorize and assign roles to mobile agents, so the agent does not need to carry sensitive information (such as private keys) or perform cryptographic operations. The proposed framework is currently implemented on top of the MARISM-A project, a secure mobile agent platform based on JADE.

The use of XACML has greatly contributed to provide a uniform and standardized manner to handle distributed RBAC policies. As a side effect we have find that XACML documents are quite complex and bloated, and cannot be manually edited by administrators without a good knowledge of the language. This has forced us to create graphical user interfaces to manipulate the policy documents.

Acknowledgments

The authors would like to thank Dr. Jose Antonio Ortega-Ruiz for his comments on previous drafts of the paper and for suggesting the name XMAS.

This work has been partially funded by the Spanish Ministry of Science and Technology (MCYT) though the project TIC2003-02041.

References

1. A. Anderson, ed. Core and Hierarchical Role Based Access Control (RBAC) profile of XACML, Version 2.0. OASIS XACML-TC, Committee Draft 01, September 2004.
2. O. Bandmann, M. Dam, and B. Sadighi-Firozabadi. Constrained delegation. In *Proceedings of the IEEE Symposium on Research in Security and Privacy*, pages 131–140, Oakland, CA, May 2002. IEEE Computer Society Press.
3. M. Blaze, J. Feigenbaum, J. Ioannidis, and A. Keromytis. The KeyNote Trust Management System. RFC 2704, IETF, September 1999.
4. J.M. Bradshaw, S. Dutfield, P. Benoit, and J.D. Woolley. KAoS: Toward an industrial-strength open agent architecture. Software Agents, 1997.
5. K. Cartrysse and J.C.A. van der Lubbe. Privacy in mobile agents. In *First IEEE Symposium on Multi-Agent Security and Survivability*, 2004.
6. David W. Chadwick and Alexander Otenko. The PERMIS X.509 role based privilege management infrastructure. In *SACMAT '02: Proceedings of the seventh ACM symposium on Access control models and technologies*. ACM Press, 2002.
7. D. Chess. Security issues of mobile agents. In *Mobile Agents*, volume 1477 of *LNCS*. Springer-Verlag, 1998.
8. D. Clarke, J. Elien, C. Ellison, M. Fredette, A. Morcos, and R. Rivest. Certificate chain discovery in SPKI/SDSI. *Journal of Computer Security*, 9(9):285–322, 2001.
9. C. Ellison, B. Frantz, B. Lampson, R. Rivest, B. Thomas, and T. Ylonen. RFC 2693: SPKI certificate theory. The Internet Society, September 1999.
10. M. Erdos and S. Cantor. Shibboleth architecture v05. Internet2/MACE, May 2002.
11. D. Ferraiolo, R. Sandhu, S. Gavrila, D. Kuhn, and R Chandramouli. Proposed NIST standard for role-based access control. In *ACM Transactions on Information and System Security*, volume 4, 2001.
12. FIPA TC Ad Hoc. Fipa agent discovery service specification, November 2003.
13. JADE Board. Jade security guide. JADE-S Version 2 add-on, 2005.
14. G. Karjoth, D.B. Lange, and M. Oshima. *Mobile Agents and Security*, volume 1419 of *LNCS*, chapter A Security Model for Aglets. Springer-Verlag, 1998.
15. R. Lepro. Cardea: Dynamic access control in distributed systems. Technical report, NASA Advanced Supercomputing (NAS) Division, 2003.
16. M. Lorch, D. B. Adams, D. Kafura, M. S. R. Koneni, A. Rathi, and S. Shah. The prima system for privilege management, authorization and enforcement in grid environments. In *Fourth International Workshop on Grid Computing*, 2003.
17. G. Navarro, S. Robles, and J. Borrell. Role-based access control for e-commerce sea-of-data applications. In *Information Security Conference 2002*, September/October 2002.
18. S. Robles, J. Mir, J. Ametller, and J. Borrell. Implementation of Secure Architectures for Mobile Agents in MARISM-A. In *Fourth Int. Workshop on Mobile Agents for Telecommunication Applications*, 2002.
19. S. Cantor, J. Kemp, R. Philpott and E. Maler, ed. Assertions and Protocols for the OASIS Security Assertion Markup Language (SAML) V2.0. OASIS XACML-TC, Committee Draft 04, March 2005.
20. B. Sadighi-Firozabadi, M. Sergot, and O. Bandmann. Using authority certificates to create management structures. In *Proceedings of Security Protocols, 9th Internatinal Workshop*, April 2002.
21. N. Suri, J. Bradshaw, M. Breedya, P. Groth, G. Hill, R. Jeffers, and T. Mitrovich. An overview of the NOMADS mobile agent system. In *Proceedings of 14th European Conference on Object-Oriented Programming*, 2000.

22. T. Moses, ed. eXtensible Access Control Markup Language (XACML), Version 2.0. OASIS XACML-TC, Committee Draft 04, December 2004.

23. A. Tripathi and N. Karnik. Protected resource access for mobile agent-based distributed computing. In *Proceedings of the ICPP workshop on Wireless Networking and Mobile Computing*, 1998.

24. J. Vollbrecht, P. Calhoun, S. Farrell, L Gommans, G. Gross, B. de Bruijn, C. de Laat, M. Holdrege, and D. Spence. AAA Authorization Framework. RFC-2904, The Internet Society, August 2000.

Distributed Data Mining Protocols for Privacy: A Review of Some Recent Results[*]

Rebecca N. Wright[1], Zhiqiang Yang[1], and Sheng Zhong[2],[**]

[1] Department of Computer Science, Stevens Institute of Technology,
Hoboken, NJ 07030 USA
[2] Department of Computer Science & Engineering, State University of New York
at Buffalo, Buffalo, NY 14260 USA

Abstract. With the rapid advance of the Internet, a large amount of
sensitive data is collected, stored, and processed by different parties. Data
mining is a powerful tool that can extract knowledge from large amounts
of data. Generally, data mining requires that data be collected into a central site. However, privacy concerns may prevent different parties from
sharing their data with others. Cryptography provides extremely powerful tools which enable data sharing while protecting data privacy.

In this paper, we briefly survey four recently proposed cryptographic
techniques for protecting data privacy in distributed settings. First, we
describe a privacy-preserving technique for learning Bayesian networks
from a dataset vertically partitioned between two parties. Then, we describe three privacy-preserving data mining techniques in a fully distributed setting where each customer holds a single data record of the
database.

1 Introduction

The advances in networking, data storage, and data processing make it easy to
collect data on a large scale. Data, including sensitive data, is generally stored by
a number of entities, ranging from individuals and small businesses to national
governments. By sensitive data, we mean the data that, if used improperly, can
harm data subjects, data owners, data users, or other relevant parties. Data
mining provides the power to extract useful knowledge from large amounts of
data. However, most data mining techniques need to collect data from different
parties; in many situations, privacy concerns may prevent different parties from
sharing their data with others. An important technical challenge is how to enable
data sharing while protecting data privacy.

Data privacy is an important issue to both individuals and organizations.
Loosely speaking, data privacy means the ability to protect selected information
against selected parties. More precise definitions of data privacy have been presented in different circumstances. It is still an area of active study to determine

---

[*] This work was supported by the National Science Foundation under Grant No. CCR-
0331584.
[**] Work completed while at Stevens Institute of Technology.

the best definition of data privacy in an environment where many uses are to be enabled (some of which are unknown at the time of initial data processing) and many privacy requirements are to be met (again, some of which are unknown at the time of initial data processing).

Privacy-preserving data mining provides methods that can compute or approximate the output of a data mining algorithm without revealing at least part of the sensitive information about the data. Existing solutions can primarily be categorized into two approaches. One approach adopts cryptographic techniques to provide secure solutions in distributed settings (e.g., [LP02]). Another approach randomizes the original data so that certain underlying patterns, such as the distribution of values, are retained in the randomized data (e.g., [AS00]). Generally, the cryptographic approach can provide solutions with perfect accuracy and perfect privacy. In contrast, the randomization approach is much more efficient than the cryptographic approach, but appears to suffer a tradeoff between privacy and accuracy.

In principle, the elegant and powerful paradigm of secure multiparty computation provides general-purpose cryptographic solutions for any distributed computation [GMW87, Yao86]. However, because the inputs of data mining algorithms are huge, the overheads of the general-purpose solutions are intolerable for most applications. Instead, research in this areas seeks more efficient solutions for specific functions.

Privacy-preserving algorithms have been proposed for different data mining applications, including privacy-preserving collaborative filtering [Can02], decision trees on randomized data [AS00], association rules mining on randomized data [RH02,ESAG02], association rules mining across multiple databases [VC02,KC02], clustering [VC03,JW05,JPW06], and naive Bayes classification [KV03,VC04]. Additionally, several solutions have been proposed for privacy-preserving versions of simple primitives that are very useful for designing privacy-preserving data mining algorithms. These include finding common elements [FNP04, AES03], computing scalar products [CIK$^+$01, AD01, VC02, SWY04, FNP04, GLLM04], and computing correlation matrices [LKR03].

In this paper, we survey four of our recently proposed cryptographic privacy-preserving techniques for data mining in distributed settings [YW06, YZW05b, ZYW05, YZW05a]. Specifically, we consider two different distributed settings. In the first setting, data is distributed between two parties. The challenge is to protect data privacy while enabling the cooperation among those parties. In Section 3, we describe a privacy-preserving solution for the two parties to compute a Bayesian network on their distributed data.

In the second setting, called the *fully distributed setting*, each party holds one record of a virtual database. The fully distributed setting is particularly well suited towards the setting of mobile ad hoc networks because each party retains control of its own information. The parties can decide when they are and are not willing to participate in various data mining tasks. In this setting, we consider the scenario where a data miner wants to carry out data mining applications. The challenge is to enable the miner to learn the results of data mining tasks while protecting each party's privacy. We describe privacy-preserving solution for three tasks in the fully distributed model in Section 4.

2 Privacy Definition in Secure Multiparty Computation

In this work, we define privacy by adapting the general privacy definition in secure multiparty computation [GMW87, Yao86, Gol04]. As usual, we make the distinction between *semi-honest* and *malicious* adversaries in the distributed setting. Semi-honest adversaries only gather information and do not modify the behavior of the parties. Such adversaries often model attacks that take place after the execution of the protocol has completed. Malicious adversaries can cause the corrupted parties to execute some arbitrary, malicious operations. Here, we review the formal privacy definition with respect to semi-honest adversaries [Gol04].

Definition 1. *(privacy w.r.t semi-honest behavior)* Let $f : (x_1, \cdots, x_m) \rightarrow (y_1, \cdots, y_m)$ be an m-ary function and denote $(x_1, \cdots, x_m)$ by $\overline{x}$. For $I = \{i_1, \cdots, i_t\} \subseteq [m] = \{1, \cdots, m\}$, we let $f_I(\overline{x})$ denote $\overline{y} = \{y_{i_1}, \cdots, y_{i_t}\}$ and let $\prod$ be a m-party protocol for computing f. The view of the ith party during an execution of $\prod$ is denoted by $\mathsf{view}_i(\overline{x})$ which includes x_i, all received messages, and all internal coin flips. For $I = \{i_1, \cdots, i_t\}$, we let $\mathsf{view}_I(\overline{x}) = (\mathsf{view}_{i_1}(\overline{x}), \cdots, \mathsf{view}_{i_t}(\overline{x}))$. We say that $\prod$ privately computes F against up to t semi-honest adversaries if for all $I \subseteq \{1, \ldots, m\}$ ($|I| = t$), for all $\overline{x}$, there exists a probabilistic polynomial-time algorithm (a simulator), denoted S, such that

$$\{S((x_{i_1}, \cdots, x_{i_t}), f(\overline{x}))\} \stackrel{c}{\equiv} \{(\mathsf{view}_I(\overline{x}), OUTPUT(\overline{x}), \}$$

where $OUTPUT(\overline{x})$ denotes the output of all parties during the execution represented in $\mathsf{view}_I(\overline{x})$.

This definition asserts that the view of the parties in I can be efficiently simulated based solely on their inputs and outputs. In other words, the adversaries cannot learn anything except their inputs and final outputs. The privacy definition related with malicious adversaries can be found in [Gol04]. For two-party computation, privacy can be defined in a way slightly different from the above [Gol04].

3 Privacy-Preserving Distributed Data Mining

Cryptographic techniques provide the tools to protect data privacy by exactly allowing the desired information to be shared while concealing everything else about the data. To illustrate how to use cryptographic techniques to design privacy-preserving solutions to enable mining across distributed parties, we describe a privacy-preserving solution for a particular data mining task: learning Bayesian networks on a dataset divided among two parties who want to carry out data mining algorithms on their joint data without sharing their data directly.

3.1 Bayesian Networks

A Bayesian network (BN) is a graphical model that encodes probabilistic relationships among variables of interest [CH92]. This model can be used for data analysis and is widely used in data mining applications.

Formally, a Bayesian network for a set V of m variables is a pair (B_s, B_p). The *network structure* $B_s = (V, E)$ is a directed acyclic graph whose nodes are the set of variables. The *parameters* B_p describe local probability distributions associated with each variable. There are two important issues in using Bayesian networks: (a) Learning Bayesian networks and (b) Bayesian inferences. Learning Bayesian networks includes learning the structure and the corresponding parameters. Bayesian networks can be constructed by expert knowledge, or from a set of data, or by combining those two methods together. Here, we address the problem of privacy-preserving learning of Bayesian networks from a database vertically partitioned between two parties; in vertically partitioned data, one party holds some of the variables and the other party holds the remaining variable.

3.2 The BN Learning Protocol

A value x is *secret shared* (or simply *shared*) between two parties if the parties have values (*shares*) such that neither party knows (anything about) x, but given both parties' shares of x, it is easy to compute x. Our protocol for BN learning uses composition of privacy-preserving subprotocols in which all intermediate outputs from one subprotocol that are inputs to the next subprotocol are computed as secret shares. In this way, it can be shown that if each subprotocol is privacy-preserving, then the resulting composition is also privacy-preserving.

Our solution is a modified version of the well known *K2* protocol of Cooper and Herskovitz [CH92]. That protocol uses a score function to determine which edges to add to the network. To modify the protocol to be privacy-preserving, we seek to divide the problem into several smaller subproblems that we know how to solve in a privacy-preserving way. Specifically, noting that only the relative score values are important, we use a new score function g that approximates the relative order of the original score function. This is obtained by taking the logarithm of the original score function and dropping some lower order terms.

As a result, we are able to perform the necessary computations in a privacy-preserving way. We make use of several cryptographic subprotocols, including secure two-party computation (such as the solution of [Yao86], which we apply only on a small number of values, not on something the size of the original database), a privacy-preserving scalar product share protocol (such as the solutions described by [GLLM04]), and a privacy-preserving protocol for computing $x \ln x$ (such as [LP02]). In turn, we show how to use these to compute shares of the parameters α_{ijk} and α_{ij} that are required by the protocol.

Our overall protocol of learning BNs is described as follows. In keeping with cryptographic tradition, we call the two parties engaged in the protocol Alice and Bob.

Input: An ordered set of m nodes, an upper bound u on the number of parents for a node, both known to Alice and Bob, and a database D containing n records, vertically partitioned between Alice and Bob.

Output: Bayesian network structure B_s (whose nodes are the m input nodes, and whose edges are as defined by the values of π_i at the end of the protocol)

As the ordering of variables in V, Alice and Bob execute the following steps at each node v_i. Initially, each node has no parent. After Alice and Bob run the following steps at each node, each node has π_i as its current set of parents.

1. Alice and Bob execute privacy-preserving approximate score protocol to compute the secret shares of $g(i, \pi_i)$ and $g(i, \pi_i \cup \{z\})$ for any possible additional parent z of v_i.
2. Alice and Bob execute privacy-preserving score comparison protocol to compute which of those scores in Step 1 is maximum.
3. If $g(i, \pi_i)$ is maximum, Alice and Bob go to the next node v_{i+1} to run from Step 1 until Step 3. If one z generates the maximum score in Step 2, then z is added as the parent of v_i such that $\pi_i = \pi_i \cup \{z\}$ and Alice and Bob go back to Step 1 on the same node v_i.
4. Alice and Bob run a secure two-party computation to compute the desired parameter α_{ijk}/α_{ij}.

Further details about this protocol can be found in [YW06], where we also show how a privacy-preserving protocol to compute the parameters B_p. Experimental results addressing both the efficiency and the accuracy of the structure-learning protocol can be found in [KRWF05].

4   Privacy Protection in the Fully Distributed Setting

In this section, we consider the fully distributed setting, in which each party holds its own data record. Together these records make a "virtual database". We assume there is a data miner that wants to learn some information about this virtual database. We call each of the data-holding parties "respondents".

First, let us consider a typical scenario of mining in the fully distributed setting: the miner queries large sets of respondents, and each respondent submits her data to the miner in response. Clearly, this can be an efficient and convenient procedure, assuming the respondents are willing to submit their data. However, the respondents' willingness to submit data is affected by their privacy concerns [Cra99]. Furthermore, once a respondent submits her data to the miner, the privacy of her data is fully dependent on the miner. Because the miner is interested in obtaining a good and accurate response rate, the protection of respondents' privacy is therefore important to both the success of data mining and the respondents. By using cryptographic techniques, we describe three techniques for different mining or data collection tasks in the fully distributed setting.

4.1   Privacy-Preserving Learning Classification Model

In this section, we provide a privacy-preserving protocol to enable a data miner to learn certain classification models without collecting respondents' raw data such as to protect respondents' privacy in the fully distributed setting.

To solve this problem, we propose a simple efficient cryptographic approach which provides strong privacy for each respondent and does not give up any accuracy as the cost of privacy. The critical technique is a frequency-learning protocol that allows a data miner to compute frequencies of values or tuples of values in the respondents' data without revealing the privacy-sensitive part of the data. Unlike general-purpose cryptographic protocols, this method requires no interaction between respondents, and each respondent only needs to send a single flow of communication to the data miner. However, we are still able to ensure that nothing about the sensitive data beyond the desired frequencies is revealed to the data miner. We note that this choice of computation can itself be considered a tradeoff between privacy and utility. On one hand, the frequencies have reasonably high utility, as they can be used to enable a number of different data mining computations, but they have less privacy than requiring a different privacy-preserving computation of each kind of data mining computation the miner might later carry out with the frequencies. On the other hand, (except in degenerate cases), the frequencies have less utility than sending the raw data itself, but more privacy.

The protocol design is based on the additively homomorphic property of a variant of ElGamal encryption, which has been used in, e.g., [HS00]. The protocol itself uses the mathematical properties of exponentiation, which allows the miner to combine encrypted results received from the respondents into the desired sums.

Let G be a group where $|G| = q$ and q is a large prime, and let g be a generator of G. All computations in this section are carried out in the group G. We assume a prior set-up that results in each respondent U_i having two pairs of keys: $(x_i, X_i = g^{x_i}), (y_i, Y_i = g^{y_i})$. Define

$$X = \prod_{i=1}^{n} X_i \tag{1}$$

$$Y = \prod_{i=1}^{n} Y_i \tag{2}$$

The values x_i and y_i are private keys (i.e., each x_i and y_i is known only to respondent U_i); X_i and Y_i are public keys (i.e., they can be publicly known). In particular, the protocol requires that all respondents know the values X and Y. In addition, each respondent knows the group G and the common generator g.

In this protocol, each respondent U_i holds a Boolean value d_i, and the miner's goal is to learn $d = \sum_{i=1}^{n} d_i$. The privacy-preserving protocol for the miner to learn the frequency d is shown in Figure 1.

Using the frequency-learning protocol, we can design a privacy-preserving protocol to learn naive Bayes classifiers which are enabled solely by frequency

$$U_i \rightarrow \text{miner} : m_i = g^{d_i} \cdot X^{y_i};$$
$$h_i = Y^{x_i}.$$

$$\text{miner:} \qquad r = \prod_{i=1}^{n} \frac{m_i}{h_i};$$
$$\text{for } d = 1 \text{ to } n$$
$$\text{if } g^d = r \text{ output } d.$$

Fig. 1. Privacy-Preserving Protocol for Frequency Mining

computation. Details about this protocol can be found in [YZW05b]. To test the efficiency of the protocol, we implemented the Bayes classifier learning protocol by using OpenSSL libraries, and we ran a series of experiments in the NetBSD operating system running on an AMD Athlon 2GHz processor with 512M memory, using 512 bit cryptographic keys. Figure 2 studies how the server's (miner's) learning time changes when both the respondent number and the attribute number vary. In this experiment, we fixed the domain size of each non-class attribute to four and the domain size of the class attribute to two.

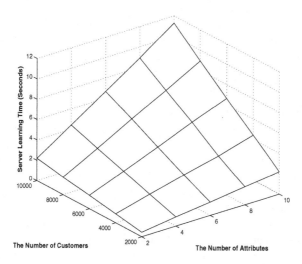

Fig. 2. Server's Learning Time for Naive Bayes Classifier vs. Number of respondents and Number of Attributes

4.2 Fully Distributed k-Anonymization

The frequency-learning protocol can be used to learn only the models which are enabled by frequency computation. However, very often a data miner wants to collect the respondents' data for the general purpose such that those data can be used for learning any model. An intuitive solution is that each respondent submits their data without any identifiers such that the miner cannot link each respondent with their submitted data and then respondents' privacy can be protected. However, even if the respondents' data do not include explicit identifiable attributes, respondents may often still be identified by using a set of attributes that act as "quasi-identifiers," e.g., {date of birth, zip code}. By using quasi-identifiers, Sweeney [Swe02b] pointed out a privacy attack in which one can find out who has what disease using a public database and voter lists.

K-*anonymization* was first proposed by Samarati and Sweeney [SS98] to address the privacy problem of quasi-identifiers. The basic idea is that a data table is k-anonymized by changing some attributes such that at least k rows have the same quasi-identifier. The existing k-anonymization methods work in the centralized setting in which the data table is located in. K-anonymization techniques include suppression and generalization methods. Suppression methods substitute the values of some attributes in quasi-identifiers with * but generalization methods substitute the values with more general ones.

K-anonymization of data can be viewed as another privacy/utility tradeoff. It publishes data that is not as useful as the original data, but that is intended to be more private. However, existing k-anonymization techniques (such as [Swe97, SS98, Sam01, Swe02b, Swe02a, MW04, BA05]) assume that the data is first available in a central location and then modified to produce k-anonymous data. In contrast, we add additional privacy protections to the k-anonymization process: distributed respondents holding their own data interact with a miner so that the miner learns a k-anonymized version of the data but no single participant, including the miner, learns extra information that could be used to link sensitive attributes to corresponding identifiers.

We give two different formulations of this problem:

- In the first formulation, given a table, the protocol needs to extract the k-*anonymous part* (i.e., the maximum subset of rows that is already k-anonymous) from it. The privacy requirement is that the sensitive attributes outside the k-anonymous part should be hidden from any individual respondent including the miner. This formulation is only suitable if the original table is already close to k-anonymous, as otherwise the utility of the result will be significantly reduced.
- In the second formulation, given a table, the protocol needs to suppress some entries of the quasi-identifier attributes, so that the entire table is k-anonymized. The privacy requirement is that the suppressed entries should be hidden from any individual participant. This formulation is suitable even if the original table is not close to k-anonymous.

In [ZYW05], we present efficient solutions to both formulations. Our solutions use cryptography to obtain provable guarantees of their privacy properties, relative to standard cryptographic assumptions. Our solution to the first problem formulation does not reveal any information about the sensitive attributes outside the k-anonymous part. Our solution to the second problem formulation is not fully private, in that it reveals the k-anonymous result as well as the distances between each pair of rows in the original table. We prove that it does not reveal any additional information. Our protocols enhance the privacy of

k-anonymization by maintaining end-to-end privacy from the original data to the final k-anonymous results.

4.3   Anonymity-Preserving Data Collection

We next consider another task in the fully distributed setting, which can again be considered as different point on the utility/privacy tradeoff. This task is suitable for data collection when the data *is* considered to provide sufficient privacy as long as it can be collected anonymously (i.e., without the data collector learning which data belongs to which respondent). An example of this scenario might be if the miner is a medical researcher who studies the relationship between dining habits and a certain disease. Because a respondent does not want to reveal what food she eats and/or whether she has that disease, she may give false information or decline to provide information. However, even if each respondent's data does not contain any identifiable attribute, the privacy of each respondent cannot be guaranteed because the miner can link the respondent's identity with their submitted data through the communication channel, e.g., by IP address. One possible solution is that the miner collects data *anonymously*. That is, he collects records from the respondents containing each respondent's dining habits and health information related to that disease, but does not know which record came from which respondent. In some settings, this idea that a response is "hidden" among many peers is enough to make participants respond.

We generalize this idea to propose an approach called anonymity-preserving data collection. Specifically, we propose that the miner should collect data in such a way that he is unable to link any piece of data collected to the respondent who provided that piece of data. In this way, respondents do not need to worry about their privacy. Furthermore, the collected data is not modified in any way, and thus the miner will have the freedom to apply any suitable mining algorithms to the data. As discussed above, this is therefore only useful for providing privacy if each respondent's data does not contains identifiable attributes and if the responses themselves do not provide too many clues to the respondent's identity.

We summarize our protocol here. Respondents are divided into many smaller groups of size N, in which the respondents' data are denoted by $(d_1, \ldots, d_N)$. A larger N will provide more anonymity but less efficiency, and vice versa. Our goal is that the miner should obtain a random permutation of the respondents' data $(d_1, \ldots, d_N)$, without knowing which piece of data comes from which respondent. To achieve this goal, we use ElGamal encryption together with a *rerandomization* technique and a *joint decryption* technique. In the ElGamal encryption scheme, one cleartext has many possible encryptions, as the random number r can take on many different values. ElGamal supports rerandomization, which means computing a different encryption of M from a given encryption of M. A related operation is permutation of the order of items, which means randomly rearranging the order of items. If we rerandomize and permute a sequence of ciphertexts, then we get another sequence of ciphertexts with the same multiset of cleartexts but in a different order. Looking at these two sequences of ciphertexts, the adversary cannot determine any information about which new ciphertext corresponds to which old ciphertext.

In our solution against semi-honest players including all respondents and the miner, t of the N respondents act as "leaders". Leaders have the special duty of anonymizing the data. At the beginning of the protocol, all respondents encrypt their data using a public key which is the product of all leaders' public keys. Note that the private key corresponding to this public key is the sum of all leaders' private keys; without the help of all leaders, nobody can decrypt any of these encryptions. The leaders then rerandomize these encryptions and permute them. Finally, the leaders jointly help the miner to decrypt the new encryptions, which are in an order independent of the original encryptions. By using digital signature and non-interactive zero-knowledge proofs, we also design the protocols against malicious miner and respondents. Further details can be found in [YZW05a].

To measure the efficiency of our protocols in practice, we implemented them using the OpenSSL libraries and measured the computational overhead. In our experiments, the length of cryptographic keys is 1024 bits. The environment used is the NetBSD operating system running on an AMD Athlon 2GHz processor with 512M memory. In the protocol against semi-honest participants, we measure the computation times of the three types of participants: regular (i.e., non-leader) respondents, leaders, and the miner. A regular respondent's computation time is always about 15ms regardless N and t. A leader's computation time is linear in N and does not depend on t. For a typical scenario where $N = 20$, the computation time of a leader is about 0.47 seconds. The miner's computation time is linear in both N and t. For a typical scenario where $N = 20$ and $t = 3$, the computation time of the miner is about 40ms. In the protocol against the malicious miner, the leader has a 10% increase over the corresponding overhead of the semi-honest protocol. The increased overhead for regular participants and the miner is negligible.

5   Discussion

We have described several privacy-preserving protocols. This remains a ripe area for research. We briefly describe some areas worthy of further investigation.

In practice, participants in a privacy-preserving protocol might behave maliciously in order to gain maximum benefits from others. Most existing work on very efficient privacy-preserving data mining, including most of ours, only provides the protocols against semi-honest adversaries. Although in principle those protocols can be modified using a general method to defend against malicious behaviors, the overhead of doing so is intolerable in practice. An important area for future research is the design of efficient mining protocols that remain secure and private even if some of the parties involved behave maliciously.

Because it aims to guarantee strong privacy for all possibilities, the general definition of privacy in secure multi-party computation is very strictly defined. Cryptographic approaches can achieve perfect privacy in principle, but one typically pays a high computational price for such privacy. For specific applications, a relaxed privacy definition might help to design efficient solutions while still be good enough to satisfy practical privacy requirements. Computing approximate mining results rather than the accurate ones might also help get the benefit of efficiency. A particularly interesting question is whether one can identify the quantitative tradeoff among efficiency, privacy, accuracy, and utility, as well as identifying solutions that achieve "good" points in that tradeoff space.

Another interesting question is how to deploy privacy-preserving techniques into practical applications. The techniques of privacy-preserving distributed data mining can be used to learn models across distributed databases. Is it feasible to define a general toolkits which are suitable for all kinds of databases with different data types? Another question is how to implement our methods without introducing covert channels to breach any party's privacy.

Particularly in the fully distributed setting, a question that remains is how to ensure either that participants provide accurate data, or that the miner can produce results in a way that is not heavily dependent on all the data being accurate. Although cryptographic techniques can force each participant to follow the protocol specifications so as to protect data privacy, but they cannot prevent participants from providing faked data to the protocols. Anonymity and privacy remove some disincentive for participants to provide fake data, but it would also be useful to design mechanisms that specifically incent participants to provide their original data.

References

[AD01]   Mikhail Atallah and Wenliang Du. Secure multi-party computational geometry. In *Proc. of the Seventh International Workshop on Algorithms and Data Structures*, pages 165–179. Springer-Verlag, 2001.

[AES03]   Rakesh Agrawal, Alexandre Evfimievski, and Ramakrishnan Srikant. Information sharing across private databases. In *Proc. of the 2003 ACM SIGMOD International Conference on Management of Data*, pages 86–97. ACM Press, 2003.

[AS00]   Rakesh Agrawal and Ramakrishnan Srikant. Privacy preserving data mining. In *Proc. of the 2000 ACM SIGMOD international conference on Management of data*, pages 439–450. ACM Press, May 2000.

[BA05]   Roberto J. Bayardo and Rakesh Agrawal. Data privacy through optimal k-anonymization. In *Proceedings of 21st International Conference on Data Engineering*, 2005.

[Can02]   John Canny. Collaborative filtering with privacy. In *Proceedings of the 2002 IEEE Symposium on Security and Privacy*, pages 45–57, Washington, DC, USA, 2002. IEEE Computer Society.

[CH92]   Greg F. Cooper and Edward Herskovits. A Bayesian method for the induction of probabilistic networks from data. *Mach. Learn.*, 9(4):309–347, 1992.

[CIK+01]   Ran Canetti, Yuval Ishai, Ravi Kumar, Michael K. Reiter, Ronitt Rubinfeld, and Rebecca N. Wright. Selective private function evaluation with applications to private statistics. In *Proc. of the 20th Annual ACM Symposium on Principles of Distributed Computing*, pages 293–304. ACM Press, 2001.

[Cra99]   Lorrie F. Cranor. Special issue on internet privacy. *Communications of the ACM*, 42(2), 1999.

[ESAG02]   Alexandre Evfimievski, Ramakrishnan Srikant, Rakesh Agrawal, and Johannes Gehrke. Privacy preserving mining of association rules. In *Proc. of the Eighth ACM SIGKDD International Conference on Knowledge Discovery and Data Mining*, pages 217–228. ACM Press, 2002.

[FNP04]   Michael J. Freedman, Kobbi Nissim, and Benny Pinkas. Efficient private matching and set intersection. In *Advances in Cryptology – EUROCRYPT 2004*, volume 3027 of *LNCS*, pages 1–19. Springer-Verlag, 2004.

[GLLM04]   Bart Goethals, Sven Laur, Helger Lipmaa, and Taneli Mielikäinen. On private scalar product computation for privacy-preserving data mining. In *Proc. of the Seventh Annual International Conference in Information Security and Cryptology*, LNCS. Springer-Verlag, 2004. to appear.

[GMW87]   Oded Goldreich, Silvio Micali, and Avi Wigderson. How to play ANY mental game. In *Proc. of the 19th Annual ACM Conference on Theory of Computing*, pages 218–229. ACM Press, 1987.

[Gol04]   Oded Goldreich. *Foundations of Cryptography, Volume II: Basic Applications*. Cambridge University Press, 2004.

[HS00]   Martin Hirt and Kazue Sako. Efficient receipt-free voting based on homomorphic encryption. *Lecture Notes in Computer Science*, 1807:539–556, 2000.

[JPW06]   Geetha Jagannathan, Krishnan Pillaipakkamnatt, and Rebecca N. Wright. A new privacy-preserving distributed k-clustering algorithm. In *Proceedings of the Sixth SIAM International Conference on Data Mining*, 2006.

[JW05]   Geetha Jagannathan and Rebecca N. Wright. Privacy-preserving distributed k-means clustering over arbitrarily partitioned data. In *Proc. of the 11th ACM SIGKDD International Conference on Knowledge Discovery and Data Mining*, pages 593–599. ACM Press, 2005.

[KC02]   Murat Kantarcioglu and Chris Clifton. Privacy-preserving distributed mining of association rules on horizontally partitioned data. In *Proc. of the ACM SIGMOD Workshop on Research Issues on Data Mining and Knowledge Discovery (DMKD'02)*, pages 24–31, June 2002.

[KRWF05]   Onur Kardes, Raphael S. Ryger, Rebecca N. Wright, and Joan Feigenbaum. Implementing privacy-preserving Bayesian-net discovery for vertically partitioned data. In *Proceedings of the ICDM Workshop on Privacy and Security Aspects of Data Mining*, Houston, TX, 2005.

[KV03]   Murat Kantarcioglu and Jaideep Vaidya. Privacy preserving naive Bayes classifier for horizontally partitioned data. In *IEEE Workshop on Privacy Preserving Data Mining*, 2003.

[LKR03]   Kun Liu, Hillol Kargupta, and Jessica Ryan. Multiplicative noise, random projection, and privacy preserving data mining from distributed multiparty data. Technical Report TR-CS-03-24, Computer Science and Electrical Engineering Department, University of Maryland, Baltimore County, 2003.

[LP02]   Yehuda Lindell and Benny Pinkas. Privacy preserving data mining. *J. Cryptology*, 15(3):177–206, 2002.

[MW04]   Adam Meyerson and Ryan Williams. On the complexity of optimal k-anonymity. In *Proc. 22nd ACM SIGMOD-SIGACT-SIGART Symposium on Principles of Database Systems*, Paris, France, June 2004.

[RH02]   Shariq Rizvi and Jayant R. Haritsa. Maintaining data privacy in association rule mining. In *Proc. of the 28th VLDB Conference*, 2002.

[Sam01]   Pierangela Samarati.   Protecting respondent's privacy in microdata
            release.   *IEEE Transactions on Knowledge and Data Engineering*,
            13(6):1010–1027, 2001.
[SS98]   Pierangela Samarati and Latanya Sweeney. Generalizing data to provide
            anonymity when disclosing information (abstract). In *Proc. of the 17th
            ACM SIGACT-SIGMOD-SIGART Symposium on Principles of Database
            Systems*, page 188. ACM Press, 1998.
[Swe97]   Latanya Sweeney. Guaranteeing anonymity when sharing medical data,
            the datafly system. *Journal of the American Medical Informatics Associ-
            ation*, 1997.
[Swe02a]   Latanya Sweeney. Achieving k-anonymity privacy protection using gener-
            alization and suppression. *International Journal of Uncertainty, Fuzziness
            Knowledge-Based Systems*, 10(5):571–588, 2002.
[Swe02b]   Latanya Sweeney.   k-anonymity: a model for protecting privacy.   *In-
            ternational Journal of Uncertainty, Fuzziness Knowledge-Based Systems*,
            10(5):557–570, 2002.
[SWY04]   Hiranmayee Subramaniam, Rebecca N. Wright, and Zhiqiang Yang.
            Experimental analysis of privacy-preserving statistics computation. In
            Proc. of the VLDB Worshop on Secure Data Management, pages 55–66,
            August 2004.
[VC02]   Jaideep Vaidya and Chris Clifton. Privacy preserving association rule min-
            ing in vertically partitioned data. In *Proc. of the Eighth ACM SIGKDD
            International Conference on Knowledge Discovery and Data Mining*, pages
            639–644. ACM Press, 2002.
[VC03]   Jaideep Vaidya and Chris Clifton. Privacy-preserving k-means clustering
            over vertically partitioned data. In *Proc. of the Ninth ACM SIGKDD
            International Conference on Knowledge Discovery and Data Mining*, pages
            206–215. ACM Press, 2003.
[VC04]   Jaideep Vaidya and Chris Clifton. Privacy preserving naive Bayes classifier
            on vertically partitioned data. In *2004 SIAM International Conference on
            Data Mining*, 2004.
[Yao86]   Andrew C.-C. Yao. How to generate and exchange secrets. In *Proc. of
            the 27th IEEE Symposium on Foundations of Computer Science*, pages
            162–167, 1986.
[YW06]   Zhiqiang Yang and Rebecca N. Wright. Privacy-preserving Bayesian net-
            work computation on vertically partitioned data. In *IEEE Transactions
            on Knowledge and Data Engineering*, 2006. to appear.
[YZW05a]   Zhiqiang Yang, Sheng Zhong, and Rebecca N. Wright.   Anonymity-
            preserving data collection. In *Proceedings of the 11th ACM SIGKDD
            International Conference on Knowledge Discovery and Data Mining*, 2005.
[YZW05b]   Zhiqiang Yang, Sheng Zhong, and Rebecca N. Wright. Privacy-preserving
            classification of customer data without loss of accuracy. In *Proceedings of
            the 2005 SIAM International Conference on Data Mining*, 2005.
[ZYW05]   Sheng Zhong, Zhiqiang Yang, and Rebecca N. Wright. Privacy-enhancing
            k-anonymization of customer data. In *Proceedings of the 24th ACM Sym-
            posium on Principles of Database Systems*, 2005.

Detecting Impersonation Attacks in Future Wireless and Mobile Networks[*]

Michel Barbeau, Jyanthi Hall, and Evangelos Kranakis

School of Computer Science, Carleton University, Ottawa, K1S 5B6, Canada

Abstract. Impersonation attacks in wireless and mobile networks by professional criminal groups are becoming more sophisticated. We confirm with simple risk analysis that impersonation attacks offer attractive incentives to malicious criminals and should therefore be given highest priority in research studies. We also survey our recent investigations on Radio Frequency Fingerprinting and User Mobility Profiles and discuss details of our methodologies for building enhanced intrusion detection systems for future wireless and mobile networks.

1 Introduction

As wireless systems are increasingly being used for critical communication it is becoming a challenge to keep electronic data transmissions secure. In general, it is difficult to implement effective security in small-footprint devices having low processing power, low memory capacity and using unreliable, low bandwidth. It is proving challenging to adapt wire-line technologies to the constrained mobile/wireless environment, enforce backward compatibility, and take account of heterogeneity.

Existing wire-line intrusion detection systems (IDSs) are classified either by the data collection mechanism (host-based, network-based), or by the detection technique (signature-based, anomaly-based, specification-based). No such simple classification is possible in wireless systems which are characterized by unavailability of key traffic concentration points, impossibility to rely on a centralized server, difficulty to secure signature distribution, and possible presence of *rogue* hosts.

Enabling wireless technologies like WTLS (Wireless Transport Layered Security) within WAP (Wireless Application Protocol), WEP (Wired Equivalent Privacy), TKIP (Temporal Key Integrity Protocol), Counter Mode CBC-MAC, Wireless PKI, Smart Cards, offer security with various degrees of success. Nevertheless, wireless devices (smart phones, PDAs, etc.) with Internet connectivity are becoming easy targets of malicious code (Cabir, Skulls, Mquito, Wince.Duts, Metal Gear, Lasco, Gavno, etc.) The question arising is *why can we not merely adapt methods from wire-line security?* We cannot, because wireless security is

---

[*] Research supported in part by NSERC (Natural Sciences and Engineering Research Council of Canada) and MITACS (Mathematics of Information Technology and Complex Systems) grants.

M. Burmester and A. Yasinsac (Eds.): MADNES 2005, LNCS 4074, pp. 80–95, 2006.

different from wire-line security. In fact wireless networks lack appropriate security infrastructure, and give potential attackers easy transport medium access. Rogue wireless access points deserve particular attention since they are not authorized for operation. They are usually installed either by employees (that do not understand security issues) or by hackers (to provide interface to a corporate network). Attention has been paid to finding rogues by using 1) wireless sniffing tools (e.g., AirMagnet or NetStumber), walking through facilities and looking for access points that have authorized Medium Access Control (MAC) addresses, vendor name, or security configurations, 2) a central console attached to the wired side of the network for monitoring (e.g., AirWave), 3) a free Transmission Control Protocol (TCP) port scanner (e.g., SuperScan 3.0), that identifies enabled TCP ports. However, are these techniques effective?

Attacks can be undertaken from an *armchair* or *war-walking* or even *war-driving*. Malicious attackers can be divided into two types. 1) *Focused attackers:* these are full time, dedicated professionals who have nothing better to do than target a specific enterprise. 2) *Opportunistic attackers:* that will attack a wireless network because it is there (a target of opportunity with no functional level of security that can be easily compromised). Although several attacks have been addressed including active/passive eavesdropping, man-in-the-middle, replay (including de-authentication and de-association), session hijacking, using traffic analysis, and masquerading, existing authentication schemes cannot fully protect hosts from well-known impersonation attacks.

1.1   Outline of the Paper

In this paper, first we confirm in Section 2 with simple risk analysis that impersonation attacks in wireless and mobile networks offer strong incentives to malicious criminal groups and should therefore be given highest priority in research studies. In Section 3, we survey our recent investigations on Radio Frequency Fingerprinting and User Mobility Profiling and discuss details of our methodologies for building enhanced intrusion detection systems that may prove more effective against impersonations attacks in future wireless and mobile networks.

2   Risk Analysis

An important aspect in the study of security is the understanding that not all threats are equally severe. Risk analysis enables the separation of the critical or major threats from the minor ones. Indeed, an attacker explicitly targets a wireless network only if there are valuable enough assets to pursue and payoffs are worthwhile. In understanding the risks, knowledge of the real threats helps place in context the complex landscape of security mechanisms. In this paper, we follow the risk assessment methodology by ETSI [8]. The evaluation is conducted according to three criteria: likelihood, impact and risk. The *likelihood* criterion ranks the possibility that a threat materializes as attacks. Two factors are taken into account: technical difficulties that have to be addressed by an attacker and motivation for an attacker to conduct an attack. The range of values for the likelihood is low (1), possible (2) and likely (3) respectively corresponding to a level of difficulty which is high, moderate or low or a level of motivation which is low, reasonable or high. The *impact* criterion ranks the consequences of an attack materializing a threat. The range of values for the impact is low (1), medium (2) or high (3) respectively corresponding to a threat that results in annoyance with reversible consequences or limited scope outages; loss of service for a considerable amount of time or limited financial losses; and loss of service for a long period of time, several affected users, violations of law or substantial financial losses. The likelihood and impact criteria receive numerical values from one to three (indicated between the parentheses). For a given threat, the *risk* is defined as the product of the likelihood and impact. If the numerical value of the risk is one or two, then the threat is considered to be minor and there is no need for countermeasures. If the risk is three of four, then the threat is major and needs to be handled. If the risk is six or nine, then the threat is critical and needs to be addressed in priority. We analyze hereafter the risk of impersonation in wireless networks. The results are summarized in Table 1.

2.1   Risk of Impersonation

Impersonation takes the form of device cloning, address spoofing, unauthorized access, rogue base station (or rogue access point) and replay. Device cloning consists of reprogramming a device with the hardware address of another device. This can be done also for the duration of one frame, which is an operation termed MAC address spoofing. This is a known problem in unlicensed services such as WiFi/802.11. It is an enabler for unauthorized access and various attacks such as the de-association or de-authorization attack. The problem has been under control in cellular networks. Cell phone cloning has been made illegal in many countries. It is interesting to note that a recent case of CDMA phone cloning occurred in India [17]. In WiFi/802.11 networks, the identity of a device, i.e. its hardware address, can be easily stolen over the air by intercepting frames. Presently, no wireless access technology offers perfect identity concealment over the air. Device cloning (including MAC address spoofing) is likely to occur. Some of the aforementioned attacks can cause service disruptions for considerable amounts of time. It is a threat which has at least a medium impact. There is therefore a major risk associated with the device cloning threat.

Impersonation of a legitimate user can be done to obtain unauthorized access to a wireless network. Authorization at user level has been introduced in both WiFi/802.11 [30], [5] and WiMax/802.16 [23] to mitigate the threat. A detailed analysis is conducted for WiMax. The situation is similar for WiFi. In WiMax/802.16, authorization occurs after scanning, acquisition of channel description, ranging and capability negotiation. There are three options for authorization: device list-based, X.509-based or EAP-based. If device list-based authorization is used only, then the likelihood of a subscriber impersonation attack is likely. X.509-based authorization in WiMax/802.16 uses certificates installed in devices by their manufacturers. If X.509-based authorization is used, the likelihood for a subscriber to be the victim of impersonation is possible in particular if certificates are hard coded and cannot be either renewed or revoked. The Extensible Authentication Protocol (EAP) is a generic authentication protocol [2]. EAP can be actualized with specific authentication methods such as EAP-TLS (X.509 certificate-based) [3] or EAP-SIM [14]. If EAP-based authorization is used, we believe that at this time it is safe to say that the likelihood of a subscriber impersonation attack is possible. Some of the EAP methods are being defined; security flaws are often uncovered in *unproven* mechanisms. Aboba maintains a Web page about security vulnerabilities in EAP methods [1]. It is a good idea to allow a second line of defense to play safe with EAP-based authentication. The impact of unauthorized access is medium, at least because, of the possible theft of network resources. Overall, the risk of unauthorized access in wireless networks is major or critical.

A *rogue base station* (or access point) is an attacker station that imitates a legitimate base station. The rogue base station confuses a set of subscribers (or clients) trying to get service through what they believe to be a legitimate base station. It may result in long disruptions of service. Attacks materializing this threat have high impact. The exact method of attack depends on the type of network. In a WiFi/802.11 network [29], which is carrier sense multiple access, the attacker has to capture the identity of a legitimate access point. Then it builds frames using the legitimate access point's identity. It then injects the crafted messages when the medium is available. In a WiMax/802.16 network [23], this is more difficult to do because WiMax/802.16 uses time division multiple access. The attacker must transmit while the impersonated base station is transmitting. The signal of the attacker, however, must arrive at targeted receiver subscribers with more strength and must put the signal of the impersonated base station in the background, relatively speaking. Again, the attacker has to capture the identity of a legitimate base station. Then it builds messages using the stolen identity. The attacker has to wait until time slots allocated to the impersonated base station start and transmit during these time slots. The attacker must transmit while achieving a *receive signal strength* higher than the one of the impersonated base station. The receiver subscribers reduce their gain and decode the signal of the attacker instead of the one from the impersonated base station. The rogue base station is likely to occur as there are no technical difficulties to resolve. EAP supports mutual authentication, i.e. the base station also authenticates itself to the subscriber. When EAP mutual authentication is used, the likelihood of the threat is mitigated, but not totally and remains possible for reasons similar to the ones aforementioned for EAP-based authorization. The rogue base station or access point attack is therefore a threat for which the risk is critical.

Replay protection insures that messages are freshly generated and are not retransmissions by attackers of previously intercepted messages. For the sake of efficiency, replay protection is often combined with message authentication. The first generation of WiFi/802.11 wireless networks adopted Wired Equivalent Privacy (WEP) for encryption [29]. WEP does not address either message authentication or replay protection. Recent developments, namely the WiFi

Protected Access (WPA) [5] and standard 802.11i [30], introduced much stronger confidentiality protection mechanisms in WiFi/802.11 networks. Firstly, encryption key establishment uses asymmetric key-based techniques. Secondly, WPA uses the Temporal Key Integrity Protocol (TKIP), which is RC4-based but with longer non reused keys. TKIP comprises a mechanism to insure message integrity and avoid replay, the Michael method. 802.11i supports both TKIP and Advanced Encryption Standard (AES). WiMax/802.16e uses the Data Encryption Standard (DES) or Advanced Encryption Standard (AES) to encrypt data traffic PDUs [23]. The AES includes a mechanism for the protection of integrity of data messages, their authentication and replay protection. DES does not. Replay protection of control traffic did not receive the same level of attention. In WiMax/802.16, management messages are never encrypted and not always authenticated. There are authentication mechanisms for layer management messages: the hashed message authentication code (HMAC) tuple and one-key message authentication code (OMAC) tuple. The OMAC is AES-based and includes replay protection, while to HMAC does not. The authentication mechanism for management messages to be used is negotiated at network entry. The scope of management messages to which authentication is applicable is limited in earlier versions of 802.16 (has been extended in version e). Hence, with earlier versions of 802.16 the management messages are not subject to integrity protection. Weaknesses in management messages authentication open the door to aggressions such as the man in the middle attack or rogue base station attack. The likelihood of replay attack is likely, possible or unlikely if no authentication, HMAC or OMAC is used respectively for management messages. In all cases, the impact of an attack of that type can be high because it might affect the operation of the communications. The risk is major or critical. It might be safe to allow a second line of defense against this type of attack in all the cases. Hence, it is a critical threat. The following table summarizes conclusions of our discussion.

Table 1. Risk of impersonation

Attack	Likelihood	Impact	Risk
Device cloning	Likely (3)	Medium (2)	Critical (6)
Unauthorized access			
with device list-based auth.	Likely (3)	Medium (2)	Critical (6)
with manufacturer certificate-based auth.	Possible (2)	Medium (2)	Major (4)
with EAP-based auth.	Possible (2)	Medium (2)	Major (4)
Rogue base station			
without mutual auth.	Likely (3)	High (3)	Critical (9)
with EAP-based mutual auth.	Possible (2)	High (3)	Critical (6)
Replay			
without message auth.	Likely (3)	High (3)	Critical (9)
with HMAC	Possible (2)	High (3)	Critical (6)
with OMAC	Unlikely (1)	High (3)	Major (3)

To sum up, the risk of impersonation in wireless networks is critical since the threat can be materialized into several forms of attack. Countermeasures are needed to address the threat.

3   Detecting Impersonation Attacks Using Device and User Profiles

One of the well known instantiations of identity theft, in WiFi/802.11 networks, is referred to as device cloning or Media Access Control (MAC) address spoofing. As aforementioned, this attack is carried out by obtaining the MAC address of a legitimate device, using tools that are readily available, e.g. NetStumbler [22]. This address is programmed into another device and subsequently used for obtaining unauthorized access to a Wireless Local Area Network (WLAN). Thus, the continued use of an access control list (ACL), based on MAC addresses, which are easily malleable, is no longer a viable strategy.

In order to address device cloning and MAC-address spoofing, authentication-based resolution strategies and intrusion detection-based countermeasures have been proposed. As far as resolution strategies are concerned, the use of public-key cryptography, although theoretically feasible, has some limitations. As the public/private key pair represents static data (unless it is changed periodically and that is unlikely), it can potentially be discovered using over the air and other mechanisms, especially since tamper-resistant hardware and software for hand held devices are still costly [32]. Another disadvantage [18] of this solution is the time required to manually type each MAC address and its associated public key into each access point. Unless the cost of administration is reduced via automation, this solution may not be suitable but for smaller networks. Finally, the level of resources required for public key cryptography is currently unavailable in wireless devices. Although this limitation will not persist for any length of time, as stated by Barbeau and Robert [6], even the use of elliptic key cryptography demands a level of resources that exceeds current availability.

Given these limitations and requirements, organizations may opt to address this problem using countermeasures, including intruder location by Adelstein et al. [4], commercial IDSs (e.g. AirDefense [16]) and user mobility patterns (UMPs) by Spencer [31]. Unlike the use of public-key cryptography, the use of intruder location or user mobility patterns, is less susceptible to forgery and imperson-ation attacks. For one thing, as intrusion detection mechanisms, both exploit behavioral characteristics or features, which are more difficult to forge or repli-cate. Whereas the intruder location mechanism examines the signal strength of WiFi/802.11 nodes, the use of UMPs is adopted in [31]. Second, both strategies require that an association, between a given MAC-address and its correspond-ing profile, be maintained for the purpose of detecting MAC-address spoofing. Essentially, it exemplifies the concept of using two or more pieces of identifica-tion for corroborating the identity of individuals. As far as commercial products are concerned, AirDefense does prevent MAC address spoofing by looking at the address prefix. However, this approach is limited in that the IDS makes a distinction between devices based only on the manufacturer's identification. Hence, the need to identify devices, from the *same* manufacturer, remains unfulfilled.

In light of these circumstances, there is an opportunity to further explore the use of device-based and user-based features for addressing the aforementioned problem. The application of Radio Frequency Fingerprinting (RFF) and UMPs for Anomaly-Based Intrusion Detection (ABID) is presented next. Readers are encouraged to consult the work by Hall [10] for additional details.

3.1   Radio Frequency Fingerprinting

RFF is a technology, which has been designed to capture the unique characteristics of the radio frequency energy of a transceiver, in RF-based wireless devices. Pioneered by the military to track the movement of enemy troops, it has been subsequently implemented, as an authentication mechanism, by some cellular carriers (e.g. Bell Nynex), to combat cloning fraud [24].

The key benefit of employing this technique is the increased level of difficulty, associated with the replication of a transceiverprint, i.e. a set of features extracted from the transient of a signal. As illustrated in Figure 1, the transient of a signal is associated with the start-up period of a transceiver prior to transmission. Even more importantly, it reflects the unique hardware characteristics of a transceiver. Consequently, it cannot be easily forged, unless the entire circuitry of a transceiver can be accurately replicated (e.g. requiring the theft of an authorized device).

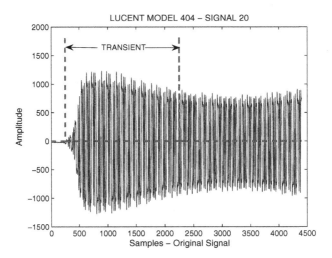

Fig. 1. Signal from a 802.11b transceiver

It is precisely this feature that has been exploited for the purpose of identifying RF-based transceivers. More specifically, a profile of each transceiver (using transceiverprints) is first created, followed by the classification of an observed transceiverprint as normal or anomalous, i.e. it does not match the transceiver profile.

3.2   RFF - Related Work

Since 1995, the level of interest in RFF continues to rise, partly motivated by the need to identify malfunctioning or illegally operated radio transmitters, in support of radio spectrum management practices. In the paper by Ellis and Serinken [7], the authors examine the amplitude and phase components of signals, captured from various transceivers (some from the same manufacturer). The general conclusion is that all transceivers do possess some consistent features (derived from amplitude and phase components), although these features may not be necessarily unique.

As far as the detection of transients is concerned, several strategies have been explored. Proposed by Shaw and Kinsner in 1997, the Threshold detection approach [28] is based on the amplitude characteristics of the signal.

Another approach, which is also based on the variance of the amplitude, is the Bayesian Step Change Detector, proposed by Ureten and Serinken [35]. Unlike the previous approach, the detection of a transient does not require the use of thresholds, i.e. it is based exclusively on the characteristics of the amplitude data. However, as the performance is less than optimal for certain types of signals, e.g. WiFi/802.11 and Bluetooth, the authors have recently proposed an enhanced detection method, referred to as the Bayesian Ramp Change Detector [27].

Finally, Hall, Barbeau and Kranakis [11] have also experimented with the use of phase characteristics of signals for detecting the start of transients. This approach can be used with WiFi/802.11 and Bluetooth signals.

In terms of classification, the use of a pattern-based classifier, such as the PNN, is advocated by many research teams including Shaw [28], Hunter [15] and Tekbas et al. [33]. The use of genetic algorithm for classification purposes has also been explored by Toonstra and Kinsner [34]. Aside from obtaining an optimal solution, this approach is rather resource-intensive. Hence, the use of genetic algorithms may not be appropriate for resource-constrained devices.

3.3   RFF - Its Use in ABID

Unlike the use of RFF for identification purposes, another option is to incorporate it into an ABID system, as illustrated by Hall, Barbeau and Kranakis [12]. The idea is to associate a MAC-address of a device with the corresponding transceiver profile. Henceforth, if an observed transceiverprint from a claimed MAC-address, matches the corresponding transceiver profile, then the MAC-address has not been spoofed.

It is generally known that current IDSs render a decision, as to whether an observed behavior/event is normal or anomalous, based on a *single* observation. In an environment that is characterized by interference and noise, delaying the decision until *multiple* observations have been classified and combined reduces the level of uncertainty. Thus, the Bayesian filter, presented by Russell and Norvig in [25], can be used to achieve this goal.

In the past, the use of static profiles has generally been the norm. However, due to factors, such as transceiver aging, there is a need to periodically capture the altered characteristics of a transceiver. Therefore, this notion of concept drift (i.e. change in behavior over time) is addressed by continuously updating the profile of a transceiver.

3.4   RFF - Evaluation

The purpose of the evaluation is two-fold: 1) to primarily assess the composition of the transceiverprint, based on the false alarm and detection rates and 2) to determine the impact of profile updates on these metrics.

Evaluation results for each of the 30 profiled transceivers are depicted in Figure 2. The false alarm rate (FAR), for a given transceiver, is defined as the number of reported anomalous transceiverprints divided by the total number of transceiverprints, which belong to the transceiver. On the other hand, the detection rate is similarly defined, but using the transceiverprints from the remaining transceivers. These transceivers are used for simulating intrusions. In addition, a 95% confidence interval is used for rending a classification decision, i.e. normal or anomalous.

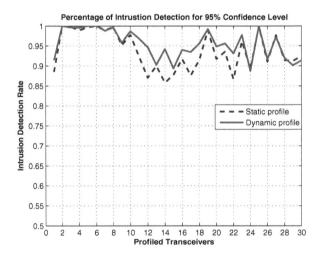

Fig. 2. Intrusion detection rate

False Alarm Rate
The FAR for this set of transceivers is 0%. Most importantly, this rate illustrates the feasibility of accurately characterizing the behavior of transceivers. Moreover, this rate is obtained when using both static and dynamic profiles (updated continuously). When a static profile is used, the FAR provides an indication as to the accuracy with which the set of transceiverprints has been selected for profiling purposes. In the case of a dynamic profile, the use of the upper/lower Euclidean distance thresholds and intra-transceiver variability

(i.e. the level of variability between signals from the same transceiver) permit the general characteristics of a transceiver to be preserved, without introducing abnormal behavior, e.g. outliers.

Detection Rate

The detection rate, associated with the use of static profiles, is typically lower (86-100%) for most of the transceivers, in particular transceivers 14 and 22, see Figure 2. Now, it is entirely possible that the underlying set of transceiverprints, used for profiling purposes, may not reflect the full range of variability of the corresponding transceiver. Consequently, a transceiverprint, from transceiver Y, could be mistakenly classified as belonging to transceiver X, resulting in a lower detection rate for X. This situation is remedied, to some extent, by continuously updating the profile. After a brief period of time, it begins to reflect the current behavior of the transceiver, a critical element for distinguishing between transceivers from the same manufacturer. The detection rate of 89-100% supports the use of dynamic profiles.

3.5 UMP - Related Work

In the past, UMPs have been used to address the inefficiencies of location-area based update schemes (e.g. by Wong [36] and Ma [20]) and to enhance routing in wireless mobile ad hoc networks (e.g. by Wu [37]). Their use in ABID has been investigated by Spencer [31]. Moreover, in the cellular network domain, the incorporation of user profiles into an ABID system has been evaluated by Samfat and Molva [26] as well as by Sun and Yu [32]. Samfat and Molva have also studied the use of usage patterns in anomaly detection. The novelty of their approach is that the detection procedure is carried out in *real-time*, i.e. within the duration of a typical call. Sun and Yu propose an *on-line* anomaly detection algorithm where the key distinguishing characteristic is the use of sequences of cell IDs traversed by a user. Both approaches do take into consideration the need for addressing concept drift. These solutions specifically target *phone* theft. It is not surprising that they leverage the existing infrastructure of cellular networks. A common characteristic of these solutions is the use of simulated data for both profiling and classification purposes. In our opinion, what would prove useful for addressing not only device cloning and MAC-address spoofing, but impersonation attacks in general is: *A generic user-based IDS mechanism.*

3.6 UMP - Its Use in ABID

We review hereafter our experience on the use of UMPs for ABID purposes. Our work considers a number of distinguishing features. Firstly, as far as the user profiles are concerned, our work is based on real mobility data collected as location broadcasts (LBs). The LBs contain latitude and longitude coordinates (LCs) and other related data. They were captured using the Automatic Position Reporting System (APRS). APRS is a packet radio-based system for tracking mobile objects. It captures and reports on locations, weather and other information for a geographical area, e.g. country or city. A detailed discussion of the APRS architecture is provided by Filjar and Desic [9].

With respect to classification, we use an Instance-Based Learning (IBL) classifier [19]. It compares an observed *set* of mobility sequences of a user to the training patterns in his/her profile. As with RFF, a set of mobility sequences, rather than a single sequence, is used to accommodate a moderate level of deviation in behavior. For a given user, if the Noise Suppressed Similarity Measure to Profile (NSMP) value, an average similarity measure formally defined in [19], falls within pre-established minimum and maximum thresholds (or acceptance region), then mobility sequences are considered normal. Otherwise, an alert is generated. The technical details of this approach are available in a companion paper [13].

3.7   UMPs - Evaluation

We discuss our evaluation of the use of UMPs and IBL for ABID. An objective is to determine the correlation between different precision levels (PLs) used for characterization and resulting false alarm and detection rates. A PL refers to a level of granularity for LCs, i.e. the number of decimals used to represent the latitude and longitude of every coordinate. PLs corresponding to one, two and three decimals are used in this study. The intra-user variability, an undesirable feature, increases with the PL.

It has been suggested by Markoulidakis [21] that nearly 50% of all mobile users of public transportation, e.g. buses, can be characterized. This statistic has been confirmed to some extent by Wu [37]. Users who took busses in the area of Los Angeles are the objects of our study. Los Angeles was selected because of the high density of APRS users. The top 50 users (those who had transmitted the highest number of LBs) were selected to participate in the study.

The evaluation was carried out for each of the 50 profiled users. For each user, the mobility sequences, which were created using the LBs, were divided into training, parameter and test data. The user-based thresholds were established by comparing the sequences in the parameter data to the patterns in the training data. In order to determine the percentage of false alarms (FAs), a comparison was made between the sequences in the test data of the user and his/her training patterns. The resulting NSMP values, which fell outside the acceptance region, were considered FAs. On the other hand, the detection rate or true detect (TDs) was obtained by comparing the test sequences of the remaining users to the training patterns of the user being evaluated. As with FAs, all NSMP values, outside the acceptance region, were considered TDs. Statistics, corresponding to these metrics, were obtained for all profiled users.

In order to simplify the analysis and subsequent discussion of results, three classes of users were defined. Class one (40% of the users) represents users who exhibit consistent Behavior. Class two (56%) and three (4%) are associated with users having progressively more chaotic behavior. We focus on the results obtained for representatives from each class, namely users that we number 19, 23 and 41 respectively in classes one, two and three.

False Alarm and Detection Rates

Figure 4 illustrates the percentage of FAs and TDs corresponding to each of three PLs used. We begin by analyzing the results for user 19. We observe that there are no FAs for all three PLs. As illustrated in Figure 3, the minimum threshold, associated with a given PL, shifts towards the lower end of the spectrum, as the PL is increased, e.g. from PL 2 to PL 3. However, all three of them (e.g. 2, 5 and 16) are greater than the value of zero. It is an indication that the mobility sequences, in the parameter data, are similar to those in the training data. Furthermore, the mobility sequences of the test data are also similar to the parameter data, which had been used to establish the thresholds.

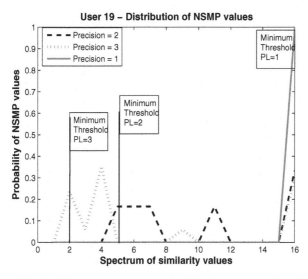

Fig. 3. Characterization using different precision levels

The TDs decrease as the PL is increased. Further scrutiny reveals that this behavior is also appropriate, given the impact of a PL on the NSMP distribution. Therefore, as the minimum thresholds shift towards the lower end, the probability of classifying intrusions as normal behavior becomes higher. This results in a decrease in the TD rate.

The characterization of user 23, on the other hand, is not as optimal. The minimum threshold of value zero is an indication that there are sequences in the parameter data, which are absent in the training data. Nevertheless, the test sequences are similar to those in the parameter set, resulting in zero FAs. In addition, the value of the minimum threshold, have also permitted all intrusions to remain undetected, resulting in a TD rate of zero. As the PL is increased to two and the maximum threshold becomes equivalent to the minimum threshold, it becomes more evident that the test sequences are dissimilar to those in the parameter data. However, they are similar to the training patterns. Consequently, the FA rate becomes 100%. The corresponding TD rate, at PL2, also increases due to the fact that the intrusions, which had fallen outside the minimum and maximum thresholds of zero, are now being detected at this level. Finally, as the PL is increased to three, the number of FAs decreases, as a result of the increase in intra-user variability between the test sequences and the training patterns. As expected, the TD rate also decreases as the PL is increased. Simply stated, the increase in inter-user variability, in conjunction with the pre-established thresholds, has influenced the detection rate of intrusions.

Finally, results for user 41 are very interesting, although somewhat misleading. We observe that, as with user 19, there are zero FAs for all three PLs. However, unlike user 19, the minimum and maximum thresholds of zero and four respectively, for all PLs, have permitted the NSMP values of all test sequences to fall within the narrow acceptance region. Similarly, the minimum threshold of value zero has also prevented all intrusions from being detected, even when the test sequences of all other users are dissimilar to the training patterns of user 41.

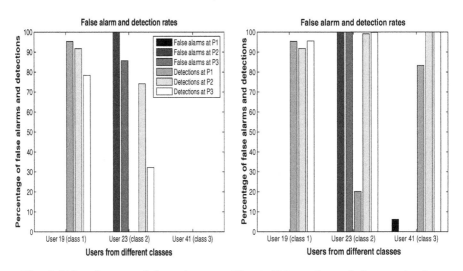

Fig. 4. False alarms and detections   **Fig. 5.** Using enhanced characterization

Enhanced Characterization

What can be ascertained, from the previous evaluation exercise, is the need to improve characterization, i.e. shift the minimum threshold to a value greater than zero. One simple strategy is to incorporate the mobility sequences from the parameter data, which have a NSMP value of zero, into the training data.

Figure 5 demonstrates the application of this strategy and the resulting impact on FA and TD rates. With user 19, the FAs remain unchanged. The TD rates (for all PLs) have increased, as expected. Moreover, the largest increase of 19% is associated with PL 3, a desirable outcome. As far as user 23 is concerned, the three TD rates, associated with PL 1, PL 2 and PL 3 have increased by 20%, 33% and 23% respectively. However, the FAs for PL 3 has also increased due to the dissimilarity of some of the test sequences to those in the parameter set. Finally, the results for user 41 exemplify the potential benefit of this strategy. Although a 5% increase in the FAs (at PL 1) has been incurred, there is, nevertheless, a significant improvement in the TDs (85%, 100%, 100%), associated with the three PLs.

4   Conclusion

Using simple risk analysis, it can be demonstrated that existing authentication schemes cannot fully protect hosts in a wireless network from impersonation attacks. In our research investigations, we have considered two defense strategies 1) Radio Frequency Fingerprinting, and 2) User Mobility Profiling that look promising in providing defenses against impersonation attacks in wireless and mobile networks. Further research is needed that will test their effectiveness in real-time systems and eventually integrate them into future IDSs for wireless networks.

References

1. B. Aboba. The unofficial 802.11 security web page - security vulnerabilities in EAP methods. www.drizzle.com/ aboba/IEEE/, May 2005.
2. B. Aboba, L. Blunk, J. Vollbrecht, J. Carlson, and H. Levkowetz. Extensible authentication protocol (EAP). The Internet Engineering Task Force - Request for Comments: 3748, June 2004.
3. B. Aboba and D. Simon. PPP EAP TLS authentication protocol. The Internet Engineering Task Force - Request for Comments: 2716, October 1999.
4. Frank Adelstein, Prasanth Alla, Rob Joyce, and Golden G. Richard III. Physically locating wireless intruders. In *Proceedings of the International Conference on Information Technology: Coding and Computing (ITCC'04)*, pages 482–489, 2004.
5. WiFi Alliance. Wi-fi protected access (WPA) enhanced security implementation based on ieee p802.11i standard, version 3.1, August 2004.
6. M. Barbeau and J-M. Robert. Perfect identity concealment in UMTS over radio access links. In *Proceedings of the Wireless and Mobile Computing, Networking and Communications*, Montreal, Canada, August 2005.
7. K.J. Ellis and N. Serinken. Characteristics of radio transmitter fingerprints. *Radio Science*, 36:585–597, 2001.
8. ETSI. Telecommunications and internet protocol harmonization over networks TIPHON release 4; protocol framework definition; methods and protocols for security; part 1: Threat analysis. Technical Specification ETSI TS 102 165-1 V4.1.1, 2003.
9. R. Filjar and S. Desic. Architecture of the automatic position reporting system (APRS). In *Proceedings of 46th International Symposium on Electronics in Marine (Elmar)*, page 331 335, 2004.
10. J. Hall. *Anomaly-based Intrusion Detection in Wireless Networks using Device and User-based Profiles*. PhD thesis, Carleton University, Fall 2005.
11. J. Hall, M. Barbeau, and E. Kranakis. Detection of Transient in Radio Frequency Fingerprinting using Signal Phase. In *proceedings of the 3rd IASTED International Conference on Wireless and Optical Communications (WOC 2003)*, pages 13–18, Banff, Canada, July 2003. ACTA Press.
12. J. Hall, M. Barbeau, and E. Kranakis. Enhancing intrusion detection in wireless networks using radio frequency fingerprinting. In *Proceedings of the 3rd IASTED International Conference on Communications, Internet and Information Technology (CIIT)*, pages 201–206, St. Thomas, U.S. Virgin Islands, November 2004.

13. J. Hall, M. Barbeau, and E. Kranakis. Using mobility profiles for anomaly-based intrusion detection in mobile networks. In *Proceedings of the Wireless and Mobile Computing, Networking and Communications*, pages 22–24, Montreal, Canada, August 2005. Preliminary version in NDSS'05 Preconference Workshop on Wireless and Mobile Security.

14. H. Haverinen and J. Salowey. Extensible authentication protocol method for GSM subscriber identity modules (EAP-SIM). Work in progress, December 2004.

15. Andrew Hunter. Feature selection using probabilistic neural networks. *Neural Computing and Applications*, 9:124–132, 2000.

16. AirDefense Inc. http://www.airdefense.net. Accessed in February 2004.

17. Financial Times Information. Mobile cloning, March 2005.

18. Alicia Laing. The Security Mechanism for IEEE 802.11 Wireless Networks. http://rr.sans.org/wireless/IEEE80211.php, 2001.

19. Terran Lane and Carla E. Brodley. Temporal sequence learning and data reduction for anomaly detection. *ACM Transactions on Information and System Security*, 2(3):295–331, August 1999.

20. W. Ma and Y. Fang. A new location management strategy based on user mobility pattern for wireless networks. In *Proceedings of the 27th Annual Conference on Local Computer Networks*, 2002.

21. J. Markoulidakis, G. Lyberopoulos, D. Tsirkas, and E. Sykas. Evaluation of location area planning scenarios in future mobile telecommunication systems. *Wireless Networks*, 1, 1995.

22. Netstumbler. http://www.netstumbler.org. Accessed in February 2004.

23. LAN MAN Standards Committee of the IEEE Computer Society, the IEEE Microwave Theory, and Techniques Society. Local and metropolitan area networks - part 16: Air interface for fixed broadband wireless access systems - amendment for physical and medium access control layers for combined fixed and mobile operation in licensed bands. Draft IEEE Standard, IEEE P802.16e/D8-2005, May 2005.

24. Michael J. Riezenman. Cellular security: better, but foes still lurk. *IEEE Spectrum*, pages 39–42, June 2000.

25. S.J. Russell and P. Norvig. *Artificial Intelligence: A Modern Approach*. Prentice Hall, 2002.

26. D. Samfat and R. Molva. IDAMN: an intrusion detection architecture for mobile networks. *IEEE Journal on Selected Areas in Communications*, 15(7):1373–1380, Sept. 1997.

27. N. Serinken and O. Ureten. Bayesian detection of Wi-Fi transmitter RF fingerprints. *Electronic Letters*, 41(6):373–374, March 2005.

28. D. Shaw and W. Kinsner. Multifractal modelling of radio transmitter transients for classification. In *Communications Power and Computing*, pages 306–312, Winnipeg Manitoba, May 1997. IEEE.

29. IEEE Computer Society. ANSI/IEEE std 802.11 - wireless LAN medium access control (MAC) and physical layer PHY specifications, 1999.

30. IEEE Computer Society. IEEE Std 802.11i-2004 IEEE standard for information technology- telecommunications and information exchange between systems- local and metropolitan area networks- specific requirements part 11: Wireless LAN medium access control (MAC) and physical layer (PHY) specifications amendment 6: Medium access control (MAC) security enhancements. Standard Number IEEE Std 802.11i-2004, 2004.

31. Jared Spencer. Use of an artificial neural network to detect anomalies in wireless device location for the purpose of intrusion detection. In *Proceedings of the IEEE*, pages 686–691, SoutheastCon, April 2005.
32. B. Sun and F. Yu. Mobility-based anomaly detection in cellular mobile networks. In *International Conference on WiSe 04*, pages 61–69, Philadelphia, Pennsylvania, USA, 2004.
33. O.H. Tekbas, O. Ureten, and N. Serinken. Improvement of transmitter identification system for low SNR transients. *Electronic Letters*, 40(3):182–183, February 2004.
34. J. Toonstra and W. Kinsner. Transient analysis and genetic algorithms for classification. In *WESCAN*. IEEE, 1995.
35. Oktay Ureten and Nur Serinken. Detection of radio transmitter turn-on transients. *Electronic Letters*, 35:1996–1997, 1999.
36. V. Wong and V. Leung. Location management for next generation personal communications networks. *IEEE Network*, pages 18–24, Sept. 2000.
37. K. Wu, J. Harms, and E.S. Elmallah. Profile-based protocols in wireless mobile ad hoc networks. *Local Computer Networks*, pages 568–575, 2001.

Anonymous Distribution of Encryption Keys in Cellular Broadcast Systems[*]

Jacek Cichoń[1], Łukasz Krzywiecki[1], Mirosław Kutyłowski[1], and Paweł Wlaź[2]

[1] Institute of Mathematics and Computer Science, Wrocław University of Technology
[2] Technical University Lublin

Abstract. We consider distribution of encryption keys for "pay-per-view" broadcasting systems with a dynamic set of users. We assume that the active recipients in such a system (i.e. those who pay for the current transmission) obtain a key necessary for decoding the transmission. If the set of recipients changes, the system has to update the key and inform the legitimate users about the change.

Communication medium we consider here is an ad hoc network of users organized in the same way as GSM or UMTS: the service area is divided into cells, each cell serves a limited number of users. Communication with the users in a cell is through a shared communication channel of this cell.

We present a procedure for distributing a new key to a new set of active users. We pursue three goals: communication volume related to a change of the encryption key should be kept as small as possible, the energy cost for each legitimate user should be low, the update process should not reveal any information about users behavior.

Our scheme is based on balanced allocation algorithms. It is simple, easy to implement, preserves anonymity. It has small communication overhead and low energy cost for the users. It works very well for a practical parameter size.

Keywords: Data broadcast, anonymity, balanced allocation.

1 Problem Statement

We consider broadcasting systems (G3 systems and the like) where an information stream is sent through a public communication channel. In order to prevent unauthorized access the transmission data is encrypted with a symmetric key and only the legitimate recipients are informed about this key.

The system might be used by a potentially large number of subscribers, each of whom shares a unique secret with the broadcasting system. It is unpredictable which subscribers and where decide to pay for the information transmitted.

Once a user decides to receive the information stream from the broadcast system, he obtains the current key used for decryption. If a user decides to log off, the billing system is informed and the key is changed so that the subscriber cannot decode the broadcast any longer. The users that stay in the system must receive a new key.

---

[*] Partially supported by KBN grant 3 T11C 011 26.

M. Burmester and A. Yasinsac (Eds.): MADNES 2005, LNCS 4074, pp. 96–109, 2006.

Communication model. We consider ad hoc systems based on cellular communication: the service area is divided into cells; each cell is served by a broadcast station sending on a channel that is accessible to all users in this cell. The users may send messages to the broadcasting system (preferably short ones due to transmission cost), while the only channel from the system to the users is through the broadcast channel.

The following complexity measures are essential for such systems:

Broadcast volume: channel capacity of the broadcast transmission is limited. So the overhead due to the key update should be kept as low as possible.

Computation complexity: the devices of the users should not perform intensive computations - we keep in mind that they should be cheap.

Energy usage: the devices of the users are often battery operated. Sending and/or receiving data is the main factor in energy consumption by mobile devices. So sending and receiving time (including the time when no message is obtained) is called energy cost and considered often as the most important complexity measure [12].

Ideally, a user who has to retrieve m information bits from the whole transmission stream inspects only the transmission value of exactly m bits. This would require full knowledge when these bits are transmitted. This is problematic, since the user does not know other users active in his service cell. Any attempt to synchronize diverse needs would require some coordination – this coordination would require energy usage! There are techniques called "air indexing" that help to find transmission time of data required [21]. Either the transmission times are explicitly given in an extra indexing data structure, or they are determined in a pseudorandom way with hashing techniques. (In fact, the approach presented in this paper is an extension of hashing techniques).

System dynamics. When a new subscriber wants to decode the transmission he must contact the broadcasting system and give it authorization for charging his account. During this interaction the current encryption key K can be transmitted to this user. There is no need to change K and inform the remaining recipients. On the other hand, if some user decides to stop receiving data, then the encryption key has to be changed and all recipients must be informed about the update.

At some moments (e.g. beginning or the end of a sport event or a movie) the set of users may change dramatically. Then it is better to distribute a new key from scratch than to use techniques for removing the users. We assume that the broadcasting system cannot predict who will be receiving the data and for how long. So we assume a scenario in which a new encryption key is transmitted periodically at fixed moments.

Privacy issues. Updating the key may leak sensitive information. First, behavior of a single user may be revealed. The second important issue is that commercially valuable information on the behavior of the subscribers community of the broadcasting system, might be revealed to competition. Using a fixed pseudonym for each user is not an appropriate solution. First, the user identity may be revealed by some usage characteristics. Second, user profiles would be shown. So more complicated "one-time" pseudonyms would be necessary.

A preferable solution would be that an external observer as well as a coalition of subscribers cannot retrieve any information on the behavior of other subscribers provided that cryptographic functions used has not been broken.

Previous results. The design of systems for distribution keys to subgroup of users has been extensively researched in recent years [8,14,18,20,22]. In [18,20] it was shown that the key tree approach reduces the server processing time complexity of group re-keying from $O(n)$ to $O(\log_d(n))$ where n is group size and d the key tree degree. This approach was proved to be optimal in [14]. The problem of transmitting a message to a small subset of users has been considered in [5].

The case considered in most papers is that the sender wishes to exclude some small number of users – *broadcast exclusion protocols* are designed. The motivation is that in this case one may gain a lot compared to a simple solution in which the sender delivers the new key to each single user that remains logged in. A broadcast exclusion protocol based on an error correcting code was proposed in [7]. Another method was proposed in [1] and independently in [9]. It uses threshold cryptosystems in order to avoid any dependence on n for the size of the private keys. There have been a lot of efforts to improve many aspects of these schemes [10,19,6]. However, many of these proposals are far from being practical at this moment – computationally intensive asymmetric methods are used. No privacy protection is implemented by these schemes.

1.1 Straightforward Approaches

Of course, one can work around the problem and establish the system in which on the side of a user there is a secure device which drops decoding the transmission in the case when the billing system stops charging the holder of the device. The problem is that the device can be reverse engineered and pirate devices may appear on the market. A more serious problem is that in the case when data services are offered by a large number of providers it will be necessary to hold many decoding devices of different providers.

We start with a discussion of ad hoc solutions that are usually considered:

Solution 1. The simplest way to distribute a new key K for users in a set $\mathcal{U}$ is to transmit a set of records, so that for each user $A \in \mathcal{U}$ one of the records is a ciphertext $E_{s(A)}(K)$, where $s(A)$ is the symmetric key shared by user A and the sender. (Asymmetric encryption would increase the size of the record and therefore it is less attractive for practical applications). The problem with this approach is that in an average case a legitimate user has to receive and decrypt half of the records to find the key - for anonymity reasons the ordering of the records has to be random. Even if there is no demand on anonymity, the set $\mathcal{U}$ is unknown to a user (the user only knows that he itself belongs to $\mathcal{U}$), so a large fraction of of ciphertexts has to be received and decrypted anyway. The problem with Solution 1 is that it has a high energy cost for the users.

Solution 2. One can facilitate the search of the appropriate ciphertext by placing an identifier in an extra data structure facilitating the search of each ciphertext. For instance, the sender may publish a random number x and then define an identifier for user A as a prefix of $H(x, A, s(A))$, for a hashing function H. The prefix should have the length such that the number of collisions (equal identifiers for two different users) is kept small. So we need at least about $\log n$ bits for each of n recipients.

A small number of collisions is not a problem. The payload transmission encrypted with the key K may be preceded by a test sequence of the form $E_K(Z)$ for some known string Z (for instance the current time). Then a user can decrypt this test message with all candidate keys and identify the right one. The size of the identifiers is a more serious issue. Even for a set of 1000 users we need at least about 10 bits. This is a large overhead compared to the ciphertext size. For 64-bit keys, this would make more than 15% overhead of transmission size. The next problem is that energy cost of a user is high - again the user has to search through the identifiers to find its own. The only good point about this solution is that the number of decryptions is reduced to just one.

1.2   Preliminaries - Lower Bound

We assume that the set of subscribers S of the broadcast system is large. The set of subscribers $\mathcal{U}$ which are in a cell considered and which are requesting the encryption key is an arbitrary subset of S of cardinality at most n. Let k denote the length of symmetric keys used, for instance $k = 64$. We assume that a user $A \in \mathcal{U}$ has a unique key $s(A)$ which is known to this user and to the broadcasting system only.

We start with the following intuitive lower bound:

Proposition 1. *Assume that if a new broadcast key K is encoded in a message T and $\mathcal{U}$ is the set of users entitled to retrieve the key K, then:*

- *each user from $\mathcal{U}$ obtains K through decoding procedure,*
- *a user that is not in $\mathcal{U}$ may retrieve K only in a negligible number of cases, i.e. in average for a fraction of $1/2^s$ of all keys.*

Assume that $|\mathcal{U}| \leq n < 2^k$ and that the encoding algorithm should work for any appointment of secret keys to the users. Then the average length of T (over the choice of keys for the users in $\mathcal{U}$) is at least

$$(k - \log n) \cdot n - f(n,k)$$

where $f(n,k) \leq (\frac{1}{2}\log n + 3) - k(1 + 2^{k-s})$.

Proof. Let $\mathcal{K}$ denote the set of keys of length k. Consider a transmission T of a new broadcast key K. We show that T, together with some additional short information, can be used to encode the subset $\mathcal{U}$ of legitimate recipients of K, where $\mathcal{U}$ is identified with the set of secret keys of the members of group $\mathcal{U}$. First consider the set $\mathcal{U}'' \subseteq \mathcal{K}$ such that $K' \in \mathcal{U}''$ if and only if T decoded with K' yields the broadcast key K. By our assumptions, $\mathcal{U} \subseteq \mathcal{U}''$ and the average number of elements in $\mathcal{U}'' \setminus \mathcal{U}$ is 2^{k-s}. Finally we encode the set $\mathcal{U}$ by the tuple (T,K,L), where $L = \mathcal{U}'' \setminus \mathcal{U}$. The average length of this tuple equals $t + k + k \cdot 2^{k-s} = t + k \cdot (1 + 2^{k-s})$, where t is the average length of T.

The number of different subsets of up to n elements from $\mathcal{K}$ is greater than $\binom{2^k}{n}$. Therefore the number of different encodings (T,K,L) is at least $\binom{2^k}{n}$. By Stirling formula we get

$$\binom{u}{v} > \frac{1}{\sqrt{2\pi}} \frac{u^{u+1/2}}{v^{v+1/2}(u-v)^{u-v+1/2}} e^{\frac{1}{12u+1} - \frac{1}{12v} - \frac{1}{12(u-v)}} .$$

So

$$\log \binom{2^k}{n} > \log 2^k \cdot (2^k + \tfrac{1}{2}) - \log n \cdot (n + \tfrac{1}{2}) - \log(2^k - n) \cdot (2^k - n + \tfrac{1}{2}) - 3$$
$$\geq n \cdot k - \log n \cdot (n + \tfrac{1}{2}) - 3 = n \cdot (k - \log n) - (\tfrac{1}{2}\log n + 3) .$$

So the average length t of T is at least $n \cdot (k - \log n) - (\frac{1}{2} \log n + 3) - k(1 + 2^{k-s})$. This completes the proof of Proposition 1.   □

The dominating term in the bound from Proposition 1 is $k \cdot n$. So even if we disregard anonymity issues the average transmission size is close to $k \cdot n$. Therefore the straight-forward Solution 1 in which we encode K separately for each user with his key cannot be improved significantly regarding communication volume (while energy cost need to be improved).

Let us examine the lower bound obtained for some interesting values of n and k. If $k = 64$, $n = 2^{10}$, and $s = 62$, then the number of bits given by the lower bound is

$$n \cdot (64 - 10) - 8 - 64(1 + 2^2) = n \cdot 54 - 328.$$

which makes about 84% of $n \cdot k$.

1.3   Key Distribution Scheme - Main Result

The main result shows that it is possible to achieve practically the same transmission size, but with strong privacy protection and with small energy cost for the users:

Algorithm 1. *We propose a scheme for transmitting a random encryption key K of length k to a set $\mathcal{U} \subseteq S$ of at most n recipients such that:*

- *any coalition $\mathcal{U}' \subseteq S$ of users (some of them receiving K) cannot say anything about the set $\mathcal{U} \setminus \mathcal{U}'$ except an upper bound on its cardinality $\min(n - u, |S \setminus \mathcal{U}'|)$, if there are u users in $\mathcal{U}'$ receiving K,*
- *every user from $\mathcal{U}$ receives certain $((n/B)d + 1) \cdot k$ bits, performs $O((n/B)d)$ arithmetic operations on k-bit numbers, and d trial decryptions,*
- *the whole message used to transmit the key has length*

$$nk \cdot \left(1 + O\left(\frac{B \ln \ln B}{d^*}\right)\right)$$

where $d^* \approx d \ln 2$,
- *the parameters B, d can be chosen freely except that $d \geq 2$.*

Even if the term on the right side of nk is not $O(1)$ (so the asymptotic behavior is worse than for the straightforward Solution 2) experiments show that the transmission volume can be bounded to size $1.01 \cdot nk$ for practical choice of parameters d, B (which is much less than in the case of Solution 2).

2   Changing the Set of Users

First we recall a widely known technique based on Lagrange interpolation of polynomials. In the second subsection we show how to reorganize transmission scheme using balanced allocations.

2.1   Anonymous Key Distribution for a Small Number of Users

We assume that there are up to n out of ω users that have to receive a new key. We fix here secure hash functions $H \neq H'$ with values almost uniformly distributed over the field $\mathbb{Z}_p = \{0, \ldots, p-1\}$, where p is a fixed prime number. We also assume that $k < \log p$ and that there is a function R such that $R : \mathbb{Z}_p \longrightarrow \{0, 1\}^k$, where each co-image $R^{-1}(x)$ has approximately the same size and both R and R^{-1} are easy to compute.

As many previous authors we use the idea of Shamir's secret sharing scheme. Let $A_{j_1}, A_{j_2}, \ldots, A_{j_n}$ be the list of users that have to receive a new key K. First the sender chooses at random a sequence q and broadcasts it to all users. Then the sender computes the values $u_i := H(q, s(A_{j_i}))$ and $x_i := H'(q, s(A_{j_i}))$ for $i = 1 \ldots, n$, where $s(A_j)$ denotes the secret assigned to the user A_j. (By a simple shifting of values we may assume that each $x_i > n$.) Then the sender determines the unique polynomial $f \in \mathbb{Z}_p[x]$ of degree n such that $f(x_i) = u_i$ for $i = 1, \ldots, n$ and $f(0) \in R^{-1}(K)$.

In the next step the sender computes values $u_{\omega+1} := f(1)$, $\ldots$, $u_{\omega+n} := f(n)$ and broadcasts them. Using Lagrange interpolation, a legitimate user A_{j_i} may reconstruct K from known values of polynomial f in $n+1$ different points, namely in $1, \ldots, n$ and in x_i. Note that the value in the last point is u_i and that it is known only to the sender and the user A_{j_i}. A user A_j, where $j \neq j_i$ for $i = 1, \ldots, n$, can also try to reconstruct f. However, using u_j leads to a false value of K, since $f(x_j) = u_j$ may occur only by accident. Since H' is like a random function, the fraction of secrets such that $f(x_j) = u_j$ should be about $1/p$.

The scheme described above protects anonymity of $A_{j_1}, A_{j_2}, \ldots, A_{j_n}$. Indeed, a user A_{j_i} cannot determine the remaining users that receive key K. User A_{j_i} may only determine the value of u_j, if A_j has to receive K and the argument x_j is known. However, there is no way to check that the values x_j, u_j are the right ones. More precisely, this would require at least finding that there is s such that $x_j = H(q, s)$ and $u_j = H'(q, s)$ for given values x_j, u_j.

The scheme described works also if the number of recipients n' is smaller than n: during the construction the sender simply fixes $n - n'$ values of f in points $1, \ldots, n - n'$ in an arbitrary way.

A major drawback of this method is that energy cost is high: a user has to receive all values $u_{\omega+1}, \ldots, u_{\omega+n}$ and q, and perform $\Omega(n)$ arithmetic operations. So the method is not practical if n is large.

2.2   Using Balanced Allocations

First we present an overview of the protocol. Let $\mathcal{U} = (A_{j_1}, \ldots, A_{j_n})$, where $n \leq \omega$, be the list of all users that are entitled to receive a new key K. The ordering of the list is set at random by the broadcaster independently for each transmission of the session key.

The basic idea is that each user is assigned to one out of B "bins". Each bin provides a message that enable some users to retrieve the key K. Such a solution must fulfill the following properties:

- each user must know from which bin he can get the key (due to energy cost),
- a user cannot deduce any significant information about bin allocation for other users,

– each bin should be responsible for about the same number of users,
– transmission for each bin should hide all information except for the key transmitted (for the users assigned to this bin).

The allocation of users to bins should work no matter what is the set of active users and without betraying any information about this set to the users. So the choice of bins for each user should be independent from the set of other users.

In order to achieve these goals we use balls-into-bins balanced allocation technique, extensively considered in literature [4,2,11,3,16,17] over a past few years. It is known that if we throw n balls into B bins independently at random and $n \geq B$, then with high probability at least one bin receive

$$n/B + \Theta\left(\sqrt{n \ln B / B}\right)$$

balls (e.g. see [13]). The general idea behind balanced allocations (so called "power-of-two-choices-principle") is to perform assignment sequentially and to give each ball a few random locations. Then the ball goes into a location that at this moment contains the smallest number of balls. If several locations have the same minimum load, then the ball is placed into an arbitrary one among them. It turns out that the result improves significantly. Recall that if d (randomly picked from B bins) locations are chosen, then the maximum load of any bin is

$$n/B + \ln \ln B / \ln d + o(1)$$

with high probability [2,3].

Our protocol is based on the balanced allocations procedure $left[d]$ from [16] (for an overview see also [17]). Let us recall it briefly. Procedure $left[d]$ splits the B bins into d groups of size B/d, the leftmost group consisting of bins 1 through B/d, the next one consisting of bins $B+1$ through $2B/d$, and so on. Procedure $left[d]$ assigns the bin sequentially one by one. For a given ball the procedure $left[d]$ chooses one bin independently and uniformly at random for each of d groups. Then the ball is placed into the least loaded of these d bins. Ties are broken by assigning the ball into the leftmost bin with the minimal number of bins among the bins considered with the minimal load.

Choice of the Bins and Encoding the Key. Let B be the number of bins. First the sender chooses a random number ρ. A preliminary assignment of the users to the bins is given by a pseudorandom function F with the range $[1,\ldots,B/d]$. Each user is assigned to d bins. For $i \in \{1,\ldots,d\}$, the ith bin chosen for the user A has index

$$b(i,A) = (i-1) \cdot B/d + F(\rho,A,s(A),i),$$

where $s(A)$ is a secret shared by A and the sender.

Then for each user $A \in \mathcal{U}$ the sender decides upon an index $c(A) \in \{1,\ldots,d\}$; in the next phase the sender constructs a message corresponding to bin $b(c(A),A)$ so that A is able to recover K from this message. The decision process $left[d]$ applied here is sequential: once the sender has already made decisions for the users $A_{j_1},\ldots,A_{j_t}$, he makes the choice for $A_{j_{t+1}}$. For this purpose he inspects all bins $b(1,A_{j_{t+1}})$, $\ldots$,

$b(d, A_{j_{t+1}})$ and select the smallest index u, denoted by $c(A_{j_{t+1}})$, such that the set $\{i \le t | b(c(A_{j_i}), A_{j_i}) = b(u, A_{j_{t+1}})\}$ has the smallest cardinality.

The following pseudocode exemplifies the procedure.

```
int b[d,A];                 // array of potential bin indexes of user 'A'
int c[A];                   // index of a bin chosen for the user
int U[A];                   // array of indexes of users to get a new key

struct s_bin                // structure describing a bin
  begin
    int Load;               // load of the bin
    int Users[n];           // array of users placed into the bin
  end

s_bin bin[B];               // array of all bins

for (i=1; i<A; i++) do      // place users from U into bins
  begin
    FindPotentialBins(p,U[i]);
    c[i]=FindLeastLoadedBin(U[i]);
    PlaceUser(U[i],c[i]);
  end

FindPotentialBins(Seed, User)
  begin
    for(i=1; i<d+1; i++)
      b[i][User]=(i-1)*B/d+F(Seed, User, s(User));
  end;

FindLeastLoadedBin(User)
  begin
    min=b[1][User];
    for(i=2; i<d+1; i++)
      if Load(b[i][User]) < Load(c[User]);
        min=b[i][User];
    return(min);
  end;

Load(BinIndex)
  begin
    return(bin[BinIndex].Load);
  end;

PlaceUser(User, BinIndex)
  begin
    bin[BinIndex].Append(User);
    bin[BinIndex].Load++;
  end;
```

Balanced allocation procedure *left*[*d*] assures that with a probability close to 1 no more than $n/B + \mu$ users are assigned to a bin for some small parameter $\mu = \ln \ln B / d^*$

(with $d^* \approx d \ln 2$). More precisely, according to [3] the maximum number of users assigned to some bin by *left*[d] is only $n/B + \frac{\ln \ln B}{d \ln \Phi_d}$ where Φ_d is expressed with use of Fibonacci numbers, i.e. $\Phi_d = \lim_{k \to \infty} \sqrt[k]{F_d(k)}$ and $F_d(k) = 0$ for $k \le 0$, $F_d(k) = 1$ for $k = 1$ and $F_d(k) = \sum_{i=1}^{d} F_d(k-1)$ for $k \ge 2$. In general $1.61 < \Phi_1 < \Phi_2 < \cdots < 2$.

Now let us explain how to encode the key K. Given a bin w the senders check for which users $A \in \mathcal{U}$ the equation $b(c(A), A) = w$ holds. If no more than $n/B + \mu$ users are assigned to a bin, then we encode K using the procedure from Section 2.1 so that exactly these users can decode K. This could be performed by the following pseudocode.

```
int Header[2*B*n+4];            // a broadcast stream header:
                                // (p,q)||(H(L),E_K(L))||(slot[1])||...)
int Points[2*n];                // values to construct the polynomial;
int slot[2*B];                  // values of the polynomial transmitted
Empty(Header);                  // initialize Header
                                // add parameters and control sequence
Header=Header||(p,q)||(H(L),E_K(L));

for(i=1; i<B+1; i++)            // for each bin
  begin
    Empty(Points);             // initialize sequence of points
    Points=Points||(0,K);      // append point (0,K)
    for(j=1; j<n+1; j++)       // for each user in a bin
        begin
          u[j]=H(q,s(bin[i][j]));
          x[j]=H'(q,s(bin[i][j]));
          Points=Points||(x[j],u[j]);
        end;

    f[i]()=Compute_Lagrange_Polynomial(Points);

    Empty(slot[i])             // initialize slot[i]

    for(k=1; k<n; k++)         // points to reconstruct polynomial f
       slot[i]=slot[i]||(k,f[i](k));
                                // append point (k,f[i](k)) to the slot
    Header=Header||(slot[i]);   // append slot[i] to the Header
  end;
```

In a rare case that some bin has more than $n/B + \mu$ users allocated the sender may choose again the parameter ρ at random and repeat the whole procedure. So the transmission size is

$$B \cdot k \cdot \left(n/B + O\left(\frac{\ln \ln B}{d \ln \Phi_d} \right) \right) = nk \cdot \left(1 + O\left(\frac{B \ln \ln B}{d \ln \Phi_d} \right) \right) .$$

Decoding by the Users. A user A determines the indices $b(1,A)$, ..., $b(d,A)$. One of them is $b(c(A),A)$, but the user A cannot determine which one. So A applies the decoding procedure described in the previous subsection for each of these bins. In this way A computes d candidate keys, one of them is K. These keys are tested against a control sequence sent - a pair $(H(L), E_K(L))$ consisting of a hash H from a random string L and its ciphertext. This could be done by the following procedure.

```
(p,q)=GetParameters(Header);
(H(L),E_K(L))=GetControlSequence(Header);

int i=0;
int u=H(q,s(A));                    // compute secret 'u'
int x=H'(q,s(A));                   // compute secret 'x'

do
  begin
    PotentialBins(p,A);             // compute potential bins 'b[d,A]'
    i++;
    slot[i]=GetSlot(b[i,A]);        // points from slot of the index 'b[i,A]'
    slot[i]=slot[i]||(x,u);         // append secret point '(x,u)'

    f[i]()=ReconstructLagrangePoynomial(slot[i]);
  end;
while not(H(L)=H(Decrypt(f[i](0),E_(L)))) // until right key K=f[i](0)
```

2.3 Features of the Protocol

General Properties. Let us discuss briefly the advantages of the protocol based on balanced allocation:

1. each user has to inspect only a fixed portion of the bits in the stream encoding the new key K, the locations from which the user retrieves the key are determined in a pseudorandom way that remain secret to everybody except the user and the sender;
2. there is no explicit addressing: each part of the stream encodes the key K;
3. the solution is scalable: the number of subscribers n might be arbitrary, increasing n implies only changing the number of bins;
4. computation overhead on the user side is negligible compared to decryption; on the sender side more computing power is necessary, but even a standard PC would make this job fast enough.

Security Features. An excluded user, say Bob, may try to derive the session key. For this purpose, Bob may try to obtain the key by reconstructing the polynomial of a single bin. Following the protocol by Bob is pointless, since the polynomial was not constructed with the pair (x, u) known by Bob, which would be used, if Bob was not excluded. Thus the Lagrangian interpolation computed by Bob yields with overwhelming probability a wrong polynomial and a wrong session key. Note that the degree of polynomials and the number of shares transmitted in a bin are such that an additional pair (argument,value) is necessary in order to determine the polynomial or, equivalently, its value for argument 0.

So, Bob has to recover the polynomial knowing only that it was constructed from pairs of the form (u, v), where $u = H(q, s)$, $v = H'(q, s)$ for some s. However, for a given pair (a, b) it is infeasible to check whether there is z such that $a = H(q, z)$, $b = H'(q, z)$.

Analysis based on different bins does not help much, since the polynomials of these bins are constructed from pairs based on different secrets.

Anonymity of the scheme is based on the following facts:

- No user identities are sent.
- Pairs $(x_i, f(x_i))$ transmitted for Lagrangian interpolation of polynomial f are such that x_i does not depend on the users assigned to the bin.
- Even if a user would learn the pairs from which a polynomial f of a bin was constructed, recovering the secrets used (and therefore the users) would require inverting hash functions. (Note that a user knows a polynomial f if and only if he knows the session key.) For the same reason the user cannot say how many of these points are really corresponding to the users that have to derive the session key.

Parameters Choice. As already mentioned [3], the maximum load of n items into B bins for $d > 1$ is at most (with high probability) $n/B + \frac{\log \log B}{d \cdot \ln \Phi_d} + O(1)$. It is worth to compare this result to the situation where $d = 1$, in which case the maximum load is $n/B + \Theta(\sqrt{n \ln B / B})$ (with high probability).

A precise estimation of the number of users in each bin for allocation procedure used is tedious, hence we provide some experimental data regarding the protocol. Let us present a summary of results. Each simulation consisted of 100 independent experiments. In the table below n denotes the number of users which receive the key, B is the number of bins, d is the number of bins assigned to a user, n/B is an average number of users allocated to a bin, $n/B + \mu$ denotes the maximal number of users assigned to a bin through $left[d]$ procedure observed in all 100 experiments. Column *max* denotes the maximal number of bins where the number of users assigned through $left[d]$ was higher than n/B.

n	B	d	n/B	$n/B + \mu$	max
10^6	10^4	1	100	145	4764
10^6	10^4	2	100	102	3109
10^6	10^4	4	100	101	1322
10^6	10^4	10	100	101	539
10^5	10^3	1	100	151	493
10^5	10^3	2	100	103	330
10^5	10^3	4	100	101	143
10^4	10^2	1	100	139	55
10^4	10^2	2	100	102	39
10^4	10^2	4	100	101	17
10^3	10	1	100	137	7
10^3	10	2	100	102	5
10^3	10	5	100	101	2
10^3	10^2	1	10	25	47
10^3	10^2	2	10	12	36
10^3	10^2	4	10	11	19

The value $d = 1$ corresponds to the standard random allocation technique. If $d > 1$, then the maximum number of users in a bin decreases quickly. It is astonishing that if $d \geq 4$, o one bin is approximately

$$n/B + 1.$$

The table shows the maximal load of a bin observed for different choices of n, B, d, obtained from experiments. Simulations prove that for a given n and B it is always possible to choose a reasonably small parameter d (typically $d = 4$) which may be applied for our purposes. The experimental data on loads of bins for different choice of n, B, d is depicted on Fig. 1.

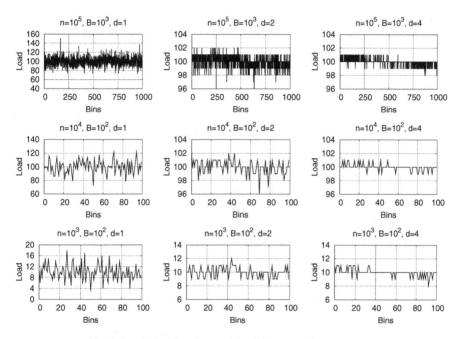

Fig. 1. Load of a bins observed for different choices of n, B, d

Let us briefly discuss the communication overhead for chosen parameters. If $n = 10^4$, $B = 10^2$ and $d = 4$ then, as the table shows, we need $10^2 + 1$ numbers of key size $k = 64$, to encode a bin. Thus the number of bits transmitted is $64 \cdot (10^2 \cdot (10^2 + 1)) = 64 \cdot (10^4 + 10^2) < (64 + 1) \cdot 10^4$. Hence the communication overhead for "addressing" is less than a bit per user. This is much less than for the simple approach discussed in Section 1.1, where transmission would consist of at least $10^6 \cdot (64 + \log(10^4)) \approx 10^6 \cdot (64 + 13)$ bits with addressing overhead of at least 13 bits per encoded key.

The choice of B should take into account the following aspects:

– too large n/B would mean a complicated reconstruction of the polynomial encoding the key - since it must have degree at least n/B,
– very small value of n/B is also inappropriate - since inevitably the load of the bins are not equal, a small increase of the number of users assigned to a bin might be a

significant change regarding multiplicative factors. For instance, if the average load is 5, then a single additional user in a bin would make the change of 20%.

3   Open Problems and Conclusions

The yet uncovered case is when the sender wishes to exclude a small number of users in an anonymous way. Of course, this could be done with the method described in Section 2.2, but we would expect much lower communication volume in this case. Let us comment that tree techniques used for excluding a small number of users are based on distributing secret keys not only to single users but also to sub-families of users. The problem in this case is how to hide who is being excluded if a specific configuration of keys is used. Another problem is that while the tree schemes reduce transmission size, the energy cost might even increase.

References

1. J. Anzai, N. Matsuzaki, and T. Matsumoto: "A Quick Group Key Distribution Scheme with Identity Revocation". ASIACRYPT'99, LNCS 1716, Springer-Verlag, 1999, pp. 333-347.
2. Y. Azar, A. Z. Broder, A. R. Karlin, and E. Upfal: "Balanced Allocations". SIAM J. Comput., 29(1), 1999, pp. 180-200.
3. P. Berenbrink, A. Czumaj, A. Steger, and B. Vöcking: "Balanced Allocations: The Heavily Loaded Case". 32nd ACM-STOC (Portland, 2000), pp. 745-754.
4. R. Cole, A. Frieze, B. M. Maggs, M. Mitzenmacher, A. W. Richa, R. K. Sitaraman, and E. Upfal: "On Balls and Bins with Deletions". RANDOM'98, LNCS 1518, Springer-Verlag, 1998, pp. 145-158.
5. A. Fiat and M. Naor: "Broadcast Encryption". CRYPTO'93, LNCS 773, Springer-Verlag, 1994, pp. 480-491.
6. Chong Hee Kim, Yong Ho Hwang, and Pil Joong Lee: "Practical Pay-TV Scheme using Traitor Tracing Scheme for Multiple Channels". Workshop on Information Security Applications (WISA) 2004, LNCS 3325, Springer-Verlag, 2004, pp. 264-277.
7. R. Kumar, S. Rajagopalan, and A. Sahai: "Coding Constructions for Blacklisting Problems without Computational Assumptions". CRYPTO'99, LNCS 1666, Springer-Verlag, 1999, pp. 609-623.
8. D. Naor, M. Naor, and J. Lotspiech: "Revocation and Tracing Schemes for Stateless Receivers". Advances in Cryptology (CRYPTO 2001), LNCS 2139, Springer-Verlag, 2001, pp. 41-62.
9. M. Naor and B. Pinkas: "Efficient Trace and Revoke Schemes". FINANCIAL CRYPTOGRAPHY 2000, LNCS 1962, Springer-Verlag, 2001, pp. 1-20.
10. N. Matsuzaki, J. Anzai, and T. Matsumoto: "Light Weight Broadcast Exclusion using Secret Sharing". 5th Australasian Conference Information Security and Privacy (ACISP2000), LNCS 1841, Springer-Verlag, 2000, pp. 313-327.
11. M. Mitzenmacher, B. Prabhakar, and D. Shah: "Load Balancing with Memory". 43rd IEEE FOCS, 2002, pp. 799-808.
12. K. Nakano and S. Olariu: "Randomized Initialization Protocols for Radio Networks". Chapter 9 in [15].
13. M. Raab and A. Steger: "Balls into bins a simple and tight analysis". Proc. of the 2nd International Workshop on Randomization and Approximation Techniques in Computer Science, Barcelona, Spain 1998, LNCS 1518, Springer-Verlag, pp. 159-170.

14. J. Snoeyink, S. Suri, and G. Varghese: "A Lower Bound for Multicast Key Distribution". Proc. of IEEE INFOCOM 2001, Anchorage, 2001, pp. 422-431.
15. I. Stojmenović (Ed.): "Handbook of Wireless Networks and Mobile Computing". Wiley 2002.
16. B. Vöcking: "How Asymmetry Helps Load Balancing". 40th IEEE-FOCS, 1999, pp. 131-140.
17. B. Vöcking: "Symmetric vs. Asymmetric Multiple-Choice Algorithms". Proc. 2nd ARACNE Workshop, Aarhus, 2001, pp. 7-15.
18. D. Wallner, E. Harder, and R. Agee: "Key Management for Multicast: Issues and Architectures". RFC 2627, June 1999.
19. Y. Watanabe, M. Numao: "Multi-round Secure Light-Weight Broadcast Exclusion Protocol with Pre-processing". ESORICS 2003, LNCS 2808, Springer-Verlag, 2003, pp. 85-99.
20. C. K. Wong, M. Goudaand, and S. Lam: "Secure Group Communications Using Key Graphs". ACM SIGCOMM'98, pp. 68-79.
21. Jianliang Xu, Dik-Lun Lee, Qinglong Hu, Wang-Chien Lee: "Data Broadcast". Chapter 10 in [15].
22. X. B. Zhang, S. S. Lam, and Dong-Young Lee: "Group Rekeying with Limited Unicast Recovery". Computer Networks, Vol. 44.6, 2004, pp. 855-870.

Non-group Cellular Automata Based One Time Password Authentication Scheme in Wireless Networks*

Jun-Cheol Jeon, Kee-Won Kim, and Kee-Young Yoo**

Department of Computer Engineering at Kyungpook National University
Daegu, Korea, 702-701
{jcjeon33, nirvana}@infosec.knu.ac.kr, yook@knu.ac.kr

Abstract. Wireless network applications mostly authenticate clients with an identity / password or pin. OTP authentication schemes have developed based on time synchronization or one-way hash functions, although they can be trouble some and they have a high computational complexity. The current paper provides secure authentication for low-power wireless devices and other applications requiring authentication that is secure against passive attacks based on replaying captured and reusable passwords. In addition, our scheme highly minimizes the computational and transmission complexity and solves time or sequence synchronization problems by applying non-group cellular automata, based on the non-reversibility and uniqueness of the state configuration.

1 Introduction

Recently, the mobile communication devices such as PDAs or smart phones have progressed rapidly. Anybody can use mobile devices to access services through the mobile network. The IEEE 802.11 wireless local area networks and mobile IP are rapidly deployed to facilitate ubiquitous communication and location independent computing in restricted spatial domains such as campuses, offices and factories [1][2].

Most wireless network applications authenticate clients with an identity / password system. Systems using reusable passwords are susceptible to attacks based on password theft. A major security concern of fixed password schemes involves eavesdropping and the subsequent replay of the password. A practical solution is a one-time password (OTP): Each password is used only once. Such schemes are safe from passive attacks of adversaries who eavesdrop and later attempt to impersonate the client [3].

The idea of OTP authentication was first proposed by Leslie Lamport [4]. Bellcore's S/KEY system, from which the OTP is derived, was proposed by Neil Haller [5]. Recently, Ben Soh and A. Joy addressed an efficient OTP scheme based

---

* This work was supported by the Brain Korea 21 Project in 2005.
** Corresponding author.

M. Burmester and A. Yasinsac (Eds.): MADNES 2005, LNCS 4074, pp. 110–116, 2006.

on a web service evaluation model [6]. All these schemes, however, have practical difficulties such as high hash overhead, additional transmission complexity, and time/sequence synchronization, since there is not an alternative solution.

A natural progression from fixed password schemes to challenge-response identification protocols may be observed by considering one-time password schemes. Variations include [3]: Shared lists of one-time passwords, sequentially updated one-time passwords, and one-time password sequences based on a one-way function. However, their drawbacks are the maintenance of the shared list, the synchronization of sequences, and the high hash overhead.

Meanwhile, Cellular automaton (CA), introduced by John Von Neumann [7], have been accepted as a good computational model for the simulation of complex physical systems, and have been used in evolutionary computations for over a decade. It can readily simulate complex growth patterns and it has also been used in various applications, such as parallel processing computations and number theory [8].

Various studies have presented the reversibility and non-reversibility of CA based on a group and non-group CA [9][10]. The group CA contains only cyclic states while the non-group CA contains both cyclic and non cyclic states. In order to satisfy the characteristics of the OTP scheme, the CA should provide the non-reversibility and uniqueness of the state configuration - that is, the previous state is not reachable from the present state of the CA, and there is no duplicate state in a CA.

Thus, this current study determines the CA which satisfies the non-reversibility and uniqueness of the state configuration, and constructs the OTP scheme based on the above characteristics. The proposed OTP scheme, based on non-group CA, eliminates the computational and transmission overhead and time/sequence synchronization problems. Also, it provides sufficient security for wireless environments.

This paper consists of five sections. In Section 2, the background of CA is illustrated briefly. Section 3 presents the proposed authentication scheme, and Section 4 provides discussion and a security analysis of our scheme. Section 5 offers concluding remarks.

2   Cellular Automata

A CA is a collection of simple cells arranged in a regular fashion. CAs can be characterized based on four properties: cellular geometry, neighborhood specification, the number of states per cell, and the rules to compute to a successor state. The next state of a CA depends on the current state and rules [7]. A CA can also be classified as linear or non-linear. If the neighborhood is only dependent on an XOR operation, the CA is linear, whereas if it is dependent on another operation, the CA is non-linear. If the neighborhood is only dependent on an EXOR or EXNOR operation, then the CA can also be referred to as an additive CA.

Among additive CAs, a CA whose dependency on neighbors is shown only in terms of XOR is called a non-complemented CA, and the corresponding rule is called the non-complemented rule. If the dependency on neighbors is shown only in terms of XNOR, the CA is called a complemented CA, and the corresponding rule is called the complemented rule. A hybrid CA can be subject to either the complemented or non-complemented rule. In addition, there are 1-dimensional, 2-dimensional, and 3-dimensional CAs according to the structure of the arrangement of cells.

A one-dimensional cellular automaton consists of a linearly connected array of n cells, each of which takes the value of 0 or 1, and an evolutionary function $f(s)$ on the state configuration, s, with q variables. The value of the cell state s_i is updated in parallel using this function in discrete time steps as $s_i{}^{t+1} = f(s_{i+j}{}^t)$ where $-r \leq j \leq r$ [11]. Table 1 shows the next states transitions by rule 90, 171, and 129.

Table 1. The states transition for 1-dimensional 2-state 3-neighborhood

111	110	101	100	011	010	001	000	Rule
0	1	0	1	1	0	1	0	90
1	0	1	0	1	0	1	1	171
1	0	0	0	0	0	0	1	129

Rule 90: $s_i{}^{t+1} = s_{i-1}(t) \oplus s_{i+1}(t)$
Rule 171: $s_i{}^{t+1} = (\neg(s_i(t) \vee s_{i-1}(t))) \oplus s_{i+1}(t)$
Rule 129: $s_i{}^{t+1} = (\neg(s_{i-1}(t) \oplus s_i(t))) \vee (s_{i-1}(t) \oplus s_{i+1}(t))$

The parameter q is usually an odd integer, i.e. $q = 2r + 1$, where r is often named the radius of the function $f(s)$; the possible configuration and the total number of rules for the radius r neighborhood are $2q$ and $2n$, where $n = 2q$. The new value of the ith cell is calculated using the value of the ith cell itself and the values of the r neighboring cells to the right and left of the ith cell. If a group rule is applied to a CA then the CA is called a group CA, otherwise the CA is a non-group CA [11].

The success of the OTP authentication to protect host systems is dependent on the non-reversibility property. CA provides both reversible and non-reversible properties by $f(s)$. In a group CA, the previous state can be easily found by computing the inverse of a rule, but it is computationally infeasible to find the inverse of a rule of a non-group CA [12].

3   Proposed Architecture Based on PBCA

The security of the CAOTP scheme is based on the non-reversibility and uniqueness of the state configuration. Such a function must be tractable in order to compute in the forward direction, but computationally infeasible to invert, and the evolved states must be distinctive. In order to achieve the conditions, a system should compute and save the non-group rule and the length according to

Table 2. The evolution sequences and length according to the given initial state (1011) by the group or non-group rule 90, 150, 171, 129

Rule ♯	Group CA		Non-group CA	
	Rule 90	Rule 150	Rule 171	Rule 129
Evolution sequence	**1011** → 0011 → 0111 → 1101 → 1100 → 1110 → **1011**	**1011** → 1000 → 1100 → 0010 → 0111 → 1010 → **1011**	1011 → 0110 → **1100** → 1001 → 0010 → **1100**	1011 → **0000** → 1111 → 0110 → **0000**
Length	6	6	5	4

the given initial state. A length represents the number of unique states in a CA. Table 2 shows the lengths when the initial state is a 4-bit vector (1011).

As shown in Table 2, the group CAs that applied rules 90 and 150 have a cyclic property whereby the initial state appears after a certain number of evolutions. The non-group CAs, that applied rules 171 and 129, do not have property as one of the previous states appears indefinitely after a certain period. It is called length, L. The proposed scheme consists of two phases: Initial phase and authentication phase. CAOPT is operated as follows.

Initial Phase
1) A client, U, chooses a pass-phrase (p) and transmits it to the base station, B.
2) The pass-phrase is concatenated with a seed (s) by B. B decides a non-group rule, f, and finds the length (L) of the non-group CA. Then, it computes and saves $\pi_0 = f^L(\pi)$, and initializes its counter for U to $C_U = 1$.
3) B transfers f, L, and π to U.

Authentication Phase
1) U computes $\pi_i = f^{L-i}(\pi)$, and transmits U, i, and $\pi_i(= f^{L-i}(\pi))$ to B.
2) B checks $i = C_U$, and $f(\pi_i) = \pi_{i-1}$. If both checks succeed, B accepts the password, sets $C_U \leftarrow C_U + 1$, and saves π_i for the next session verification.

In the initial phase (1), p may be of any length between 64 bits and 96 bits, and s should be the remainder bits, between 32 bits and 64 bits, so that the secret,π, is initialized as 128 bits length. In (2), f can be just one non-group rule or the combination of rules, e.g. $< 107, 230 >$ - That is the rules are applied to the CA alternately. The result of the concatenation, L, is passed through f. An evolutionary non-group rule function is used to define the password sequence: $f^{L-i}(\pi_0), ..., f^2(\pi_0), f(\pi_0), \pi_0$. The password for the ith identification session, $1 \le i \le L$, is defined to be $\pi_i = f^{L-i}(\pi_0)$. In the authentication phase, for the ith session, B already has the client's identity and counter, along with the verifier, π_{i-1}, so that B simply check the client's authentication by applying the evolutionary function f once.

Consider the following example. U_i chooses *'evolutionary'* as p, and sends it to B. B selects '1234' as s, and concatenated it to p. B also selects the non-group rule 107 and finds L is 882, and saves [identifier, counter, π_0]=[U_i, 1,

$f^{657}(evolutionary1234)$]. Thus, in the authentication phase, U_i computes $\pi_1 = f^{656}(evolutionary1234)$) and transmits U_i, 1, and π_1 to B. B checks $1 = C_U$ and $f(\pi_1) = f^{657}(evolutionary1234)$). After the use of all state configurations during the L times, the non-group rule f should be changed into another rule f'.

4   Discussion and Analysis

In spite of the growth of the OTP authentication schemes, previous schemes still possess several problems for wireless networks, such as high hash computations, additional communication paths, and time or sequence synchronization.

In order to solve the high hash computation problem, our scheme has taken advantage of the CA characteristics so that the computation only consists of logical bitwise operations, such as XOR, AND, and OR, while the hash function is composed of not only logical bitwise operations but also padding, and appending length. Moreover, the CA computations are performed in parallel.

CAOTP has a minimal path, only one, to check whether a client is certified in an authentication phase, and it does not need the synchronization of time. If U and B have gotten out of synchrony because of unstable computer network - U sends π_i and B uses $f(\pi_j)$ to authenticate it, with $i \neq j$ - This can be detected by repeatedly applying f to B's authenticating value until a match is obtained.

We recommend that the whole size of the state configuration, $|\pi|$ should be 128bits which consists of p and s. To reduce the risk of an exhaustive search or dictionary attacks, the client's secret pass-phrase should be between 64 bits and 96 bits. This is believed to be long enough to be secure and short enough to be entered manually. We also recommend a 3 neighbor CA which has $r = 1$ since a longer neighbor CA does not ensure better safety but only adds to the complexity of the computations.

Meanwhile, the length, L, should be reasonably long enough but not too long. If the L of a CA is too short or too long, it causes the frequent renewal of the rule or password guessing from the attacker. Thus, we recommend that the length of CA should be in the several hundreds to the several thousands. We can simply find a reasonable length by using a combination of group rules and non-group rules.

Table 3 shows the length when the rules are applied to the initial value, $evolutionary1234$. Only the CA-applied non-group rule is relatively shorter in length, 657 and 127, which correspond to rules 107 and 230. Meanwhile, the CA applied the combinations of the group rules and non-group rules $< 90, 129 >, < 171, 90 >$ which have sufficient lengths of 2,524 and 2,471. There are a number of combination rules which are sufficient in length.

The following criteria are crucial for the robust security of authentication schemes.

- Password guessing attack: It is computationally infeasible for the attacker to choose a password which is same as the current π_i from the previous session key π_{i-1}.

- Replay attack: Though the attacker replays π_i to B in the authentication phase, the request will be rejected, since π_i is used only once.

Table 3. The length corresponds to the non-group rule and combination of a group rule and non-group rule

	Non-group rule		Combination of group rule and non-group rule	
Rule ♯	Rule 107	Rule 230	Rule $< 90, 129 >$	Rule $< 171, 90 >$
Length	657	127	2524	2471

- Sever spoofing attack: The attacker can not impersonate the system, B because he/she can not compute $f^{L-i}(\pi_0)$ from π_i received from the client U (function f is not a reversible rule and π_0 is a shared secret value between U and B).

- Impersonating U: It is also infeasible that the attacker can acquire a shared secret π_0 since he cannot extract it from any information which he obtains.

- Stolen verifier attack: Though the attacker has stolen the verifier, $f^{L-i}(\pi_0)$, in ith session, $f^{L-(i+1)}(\pi_0)$ can not be computed by any method because it is infeasible to find the inverse in the non-group CA.

Thus, the proposed scheme can resist password guessing attacks, replay attacks, impersonation attacks, server spoofing attacks, and stolen verifier attacks.

5 Conclusion

One time password schemes provide significant additional protection but their use is limited due to the complexity and inconvenience regarding wireless networks. Our scheme provided not only minimal computational complexity and a transmission path but also safety guarantees. We have shown that only our scheme has logical bitwise operations and one path for authentication purposes, and it can resist the above mentioned attacks. The important aspect of this work is that we have employed an evolutionary computation based on non-group, reversible cellular automata in the OTP scheme. We believe that the CAOTP authentication scheme is suitable and practical for low power wireless network environments because of its light computations and nimble movement.

References

1. IEEE Standard 802.11. Wireless LAN medium access control (MAC) and physical layer (PHY) specification IEEE Draft Standard (1996)
2. C. Perkins.:IP mobility support. Internet Request for Comments 2002 (1996)
3. A.J.Menezes, P.C. Oorschot, S.A. Vanstone: Handbook of Applied Cryptography CRC Press (1997)
4. L. Lamport.: Password Authentication with Insecure Communication, Vol. 24 . Communication of ACM (1981) 770–772
5. N. Haller: The S/KEY One-Time Password System. University of Illinois Press, Proc. of the ISOC Symposium on Network and Distributed System Security (1994)
6. B. Soh, A. Joy: A Novel Web Security Evaluation Model for a One Time Password System. Proc. of the IEEE/WIC International Conference on Web Intelligence (2003)

7. J. Von Neumann: The theory of self-reproducing automata. University of Illinois Press, Urbana and London (1966)
8. J.C. Jeon, K. Y. Yoo: Design of Montgomery Multiplication Architecture based on Programmable Cellular Automata. Vol. 20 . Computational intelligence (2004) 495–502
9. M. Seredynski, K. Pienkosz, P. Bouvry: Reversible Cellular Automata Based Encryption. LNCS 3222, NPC 2004 (2004) 411–418
10. S. Das, B. K. Sikdar, P. Pal Chaudhuri: Charaterization of Reachable/Nonreachable Cellular Automata States. LNCS 3305, ACRI 2004. (2004) 813-822
11. C.K. Koc, A.M. Apohan: Inversion of cellular automata iterations. Vol.144 . IEE Proc. Comput. Digit. Tech. (1997) 279–284
12. M. Seredynski, P. Bouvry: Block Encryption Using Reversible Cellular Automata. LNCS 3305. (2004) 785–792

Keynote: Efficient Cryptographic Techniques for Mobile Ad-Hoc Networks

Yuliang Zheng

Department of Software and Information Systems
University of North Carolina at Charlotte
9201 University City Blvd, Charlotte, NC 28223
yzheng@uncc.edu

Abstract. The focus of this talk is on cryptographic techniques that help enhance the security of mobile ad-hoc networks while minimizing necessary overhead in terms of power consumption as well as computational and communication latency. Techniques to be discussed include both symmetric and asymmetric cryptographic schemes pertinent to the topic. Discussions will be also extended to other relevant issues, such as infrastructural support mandated by these techniques.

M. Burmester and A. Yasinsac (Eds.): MADNES 2005, LNCS 4074, p. 117, 2006.
© Springer-Verlag Berlin Heidelberg 2006

How to Generate Universally Verifiable Signatures in Ad-Hoc Networks*
(Extended Abstract)

KyungKeun Lee[1], JoongHyo Oh[2], and SangJae Moon[1]

[1] School of Electrical Engineering and Computer Science,
Kyungpook National University, Daegu, Korea
`skywalker@mail.knu.ac.kr`, `sjmoon@ee.knu.ac.kr`
[2] Korea Financial Telecommunications and Clearings
Institute, Seoul, Korea
`jhoh@kftc.or.kr`

Abstract. This paper addresses the problem of making signatures of one domain (an ad-hoc network) available in another domain (the Internet). Universal verifiability is a highly desirable property when signed documents need to be permanently non-repudiable so as to prevent dishonest signers from disavowing signatures they have produced. As a practical solution, we construct a new signature scheme where a valid signature should be generated by a couple of distinct signing keys. In the random oracle model, the signature scheme is provably secure in the sense of existential unforgeability under adaptive chosen message attacks assuming the hardness of the computational Diffie-Hellman problem in the Gap Diffie-Hellman groups.

Keywords: Ad-hoc networks, Digital signature, Diffie-Hellman problem, Bilinear map.

1 Introduction

In modern cryptography, digital signatures are among the most fundamental and prominent tools for realizing security in open distributed systems and networks. In a conventional Public Key Infrastructure (PKI) environment [ITU97], it is easy to confirm the authorship of a signed document by verifying the signature with a certificate of a given public key and, simultaneously, by checking whether the corresponding certificate is revoked, communicating with on-line Certificate Authority (CA). Within an *ad-hoc* environment, a user can also generate and verify digital signatures with the assistance of pre-established ad-hoc CA [ZH99].

Nevertheless, signatures generated from an ad-hoc network are neither easily validated nor directly compatible in any infra network since the ad-hoc network

---

* This research was supported by the MIC (Ministry of Information and Communication), Korea, under the ITRC (Information Technology Research Center) support program supervised by the IITA (Institute of Information Technology Assessment).

M. Burmester and A. Yasinsac (Eds.): MADNES 2005, LNCS 4074, pp. 118–131, 2006.

is, in general, assumed to be physically isolated from the infra network, i.e, *autonomous*. For instance, suppose that Alice is an (ad-hoc) signer and Bob is a (mobile) verifier. When Bob receives Alice's signature, he is able to verify her signature and to identify her interacting with an on-line ad-hoc CA. Now, Bob switches to an infra network such as the Internet disconnecting the ad-hoc network channel. Even if this signature was verified to be genuine before leaving the ad-hoc network, the signature is not verifiable to any other party who has received the signature from Bob within the infra network because her certificate is inapplicable outside its domain. Despite Bob's efforts to retrieve Alice's certificate status information, he will not succeed due to difficulties in communication with the ad-hoc CA. Even if the current status of her certificate can be obtained, the certificate will still be of dubious standing and even extraneous since the ad-hoc CA is definitely not a trusted authority in the new domain. Alice is then exempt from responsibility of the signature she produced. Thus, it is highly desirable to make signatures non-repudiable in order that signers should be permanently undeniable even when they are in ad-hoc networks. A trivial solution for the long-term non-repudiation is to make the signer participate in so-called a posteriori arbitration when a dispute occurs. Intuitively this requires a strong assumption that any suspicious signer would willingly take part in the arbitration; unfortunately, this assumption is not realistic.

The alternative solution proposed in this paper is to provide the *universal verifiability* for signatures. Our scheme is an a priori and non-interactive (for signers) protocol in which a mobile verifier is given an ad-hoc signature as well as a corresponding infra (actually encrypted) signature. At the same time, the verifier also obtains an additional zero-knowledge argument that demonstrates the signer's possession of (ad-hoc and infra) signing keys used for generating signatures. Interacting with a trusted "notary," the verifier will finally obtain the translation of the given ad-hoc signature while authenticating the signer's identity. This translation becomes identical to the infra signature as desired.

We construct a universally verifiable signature scheme (P2DL signature) by extending the verifiably encrypted signature scheme proposed by Boneh *et al.* [BGLS03]. We also extend the zero-knowledge proof of discrete-log equality [CP93] into a zero-knowledge proof of two discrete logarithms possession and give its formal analysis. The signature scheme presented here offers provable security against adaptive chosen message attacks under the Gap Diffie-Hellman assumption [OP01] in the random oracle model [BR93].

Motivation. Most signature schemes provide so-called "public verifiability," as long as the validation of given a message-signature pair with some redundant parameters is believed to be true. In accordance with this validation, the authenticity of corresponding signer's public (verification) key is also requisite after retrieving the signer's certificate status information from the CA according to X.509 [ITU97]. As such, the verification of signatures are on two points: the validation of given the signature, message, and parameters, plus the signer's public-key validity via revocation check. (In short, *signature validation* and *revocation*.)

In this paper, we pay attention to the verifiability of signature in two domain environment, especially in an ad-hoc network and the Internet. Ad-hoc networks are substantially autonomous. Thus, there is no communication channel from (to) an ad-hoc network to (from) the Internet, whatsoever wired or wireless connection, by definition of the ad-hoc network. *If an ordinary signature is generated from an ad-hoc network, is that signature available in the Internet?* Owing to the isolation between two networks, it looks insurmountable though.

Suppose that Alice is a signer with two signing keys $SK_{\text{ad-hoc}}$ and SK_{infra} for the purpose of signature generations in ad-hoc and infra networks, respectively. She is able to generate a signature σ under her signing key $SK_{\text{ad-hoc}}$ for the ad-hoc network and to make any user verifiable with this signature within its domain. It is always verifiable for every ad-hoc user as desired. *What if this ad-hoc signature is transferred to any other domain such as the Internet?* Even if this signature is genuine via *signature validation*, it is impossible to check whether the signer's key is revoked or not. One can ask why not use an infra signing key SK_{infra} within every ad-hoc network for the public verifiability in the Internet. An analogous problem arises for ad-hoc verifiers because of the difficulties in checking the revocation of the signer's infra network public-key. Due to the limitation of ordinary signatures in the literature, we construct a new signature scheme that offers the *universal verifiability* across two isolated domains.

Definition 1 (Universal Verifiability of Two-Domain Setting). *A signature Σ transferred from one domain is said to be universally verifiable when a given signature is publicly verifiable in both one domain $\mathcal{D}_1$ and in the other domain $\mathcal{D}_2$.*

- *(Unilaterally universal verifiability). A signature generated from $\mathcal{D}_1$ is also publicly verifiable in $\mathcal{D}_2$, but a signature generated from $\mathcal{D}_2$ is not verifiable in $\mathcal{D}_1$. In this case, the signature scheme is universally verifiable in a unilateral way from $\mathcal{D}_1$ to $\mathcal{D}_2$.*
- *(Bilaterally universal verifiability). A signature originated from any domain, $\mathcal{D}_1$ or $\mathcal{D}_2$, is publicly verifiable in both domains $\mathcal{D}_1$ and $\mathcal{D}_2$.*

Our Contributions. Supporting the unilaterally universal verifiability from an ad-hoc domain to the Internet domain, our signature scheme features several characteristics below.

- *Non-interactivity.* In the proposed scheme, signing process is not interactive. After sending a signed document to someone, the signer does not have to complete any other protocol with any party. It is highly desirable in most signature schemes and makes our scheme more practical.
- *Verifier's Mobility Support.* Due to the isolation between ad-hoc and infra networks, the only way to transfer signatures generated across domains is to allow verifiers mobile from an ad-hoc network to the Internet. We can imagine verifiers are mobile devices such as laptop computers, PDAs, or any other hand-held electrical conveniences.

- *Granularity.* In our scheme, a signer simply generates ordinary short signatures for use in each ad-hoc and infra networks. Next the signer computes the non-interactive zero knowledge (NIZK) in order to prove the signer's possession of two signing secret keys. Intuitively, the signature Σ comprises an ad-hoc signature σ_1, infra signature σ_2, and the redundancy for NIZK. Hence, after finishing appropriate validation steps, our signature Σ is no longer necessary. For ad-hoc verifiers, σ_1 is sufficient as a signature of the given message. On the other hand, σ_2 makes infra network users convinced that the given message-signature pair is authentic. In particular, the granularity feature makes our signature scheme attractive since shorter (ordinary) signature is enough for any verifier in each separated network.
- *Notarization.* A special-purpose trusted third party (TTP) of the Internet domain is assumed in our scheme. The notary's task is mainly to notify the revocation of signer's infra public key on behalf of the on-line infra CA. In addition, notary verifies any signature transferred from ad-hoc domain and then publishes a smaller ordinary signature for the Internet use only. Before notarization, this infra signature is encrypted under the notary's public key. This disables for any mobile verifier to spread signer's infra signature in any other domain without doing in the Internet. The presence of the trusted notary is not burdensome because it looks like a slightly modified CA. This also makes our protocol quite realistic.
- *Tight Security Reduction.* Interestingly, equipped with the redundant information for NIZK proof, the security of the proposed scheme is *tightly* related to the computational Diffie-Hellman problem with formal proof, while the GDH signature scheme [BLS04] adopted as ordinary signatures is *loosely* related to the same hard problem. In principle, the tight security reduction is infinitely preferable for strengthening the security of protocols.

Organization. The rest of this paper is organized as follows. The following section provides an overview of relevant background. Section 3 presents the new universally verifiable signature scheme. The security of the scheme is formally analyzed in Section 4. We finally conclude this paper in Section 5.

2   Preliminaries

2.1   Notation and Bilinear Maps

We first consider two (distinct and multiplicative) cyclic groups $\mathcal{G}_1 = \langle g_1 \rangle$ and $\mathcal{G}_2 = \langle g_2 \rangle$ of prime order q, where the discrete logarithm problem is intractable. Let ψ be a computable isomorphism from $\mathcal{G}_1$ to $\mathcal{G}_2$, with $\psi(g_1) = g_2$.

We require a bilinear map $\hat{e} : \mathcal{G}_1 \times \mathcal{G}_2 \to \mathcal{G}_T$, where the group $\mathcal{G}_T$ is multiplicative and the orders of all the given groups are the same: $\#\mathcal{G}_1 = \#\mathcal{G}_2 = \#\mathcal{G}_T = q$. A bilinear map satisfying the following properties is said to be an *admissible* bilinear map: (a) bilinear: for all $u \in \mathcal{G}_1, v \in \mathcal{G}_2$ and $a, b \in \mathbb{Z}_q, \hat{e}(u^a, v^b) = \hat{e}(u, v)^{ab}$; (b) non-degenerate: given generators g_1 and g_2, $\hat{e}(g_1, g_2) \neq 1$; (c) computable: there is an efficient algorithm to compute $\hat{e}(u, v)$ for any $u \in \mathcal{G}_1, v \in \mathcal{G}_2$.

With an isomorphism $\psi(\cdot)$, these properties lead to two more simple attributes: for any $u, w \in \mathcal{G}_1$, $\hat{e}(u, \psi(w)) = \hat{e}(w, \psi(u))$; and for any $u \in \mathcal{G}_1, v_1, v_2 \in \mathcal{G}_2$, $\hat{e}(u, v_1 v_2) = \hat{e}(u, v_1) \cdot \hat{e}(u, v_2)$. Without loss of generality, we assume these maps $\psi(\cdot)$ and $\hat{e}(\cdot)$ can be computed in one time unit for the group action, and we then call the two groups $(\mathcal{G}_1, \mathcal{G}_2)$ a bilinear group pair.

The expression $z \in_R S$ indicates that an element z is chosen randomly from the finite set S according to the uniform distribution. When parsing a string w into a sequence of fragments $U_1, U_2, \ldots, U_\ell$, we use $w \xrightarrow{\text{P}} U_1 \| U_2 \| \ldots \| U_\ell$. For simplicity, the (mod q) marker for operations on elements in $\mathbb{Z}_q$ is omitted.

2.2   Intractability Assumptions

We define several hard problems closely related to the Diffie-Hellman problem [DH76]. Consider a single cyclic group $\mathcal{G} = \langle g \rangle$, with $q = \#\mathcal{G}$ a prime.

Computational Diffie-Hellman Problem (CDH). For $a, b \in \mathbb{Z}_q$, given g, $g^a, g^b \in \mathcal{G}$, compute $g^{ab} \in \mathcal{G}$.

Decision Diffie-Hellman Problem (DDH). For $a, b, c \in \mathbb{Z}_q$, given g, g^a, g^b, $g^c \in \mathcal{G}$, answer with;
   (i) Yes, if $c = ab$. In this case, (g, g^a, g^b, g^c) is called a Diffie-Hellman quadruple.
   (ii) No, otherwise.

There is strong evidence that these two standard hard problems are closely related the hardness of computing discrete logarithms. There exist polynomial-time algorithms that either reduce the DDH to the CDH or reduce the CDH to the discrete logarithm problem. The converses of these reductions, however, are thus far not known to hold. All the three problems are widely conjectured to be intractable. A variety of cryptographic protocols have been proved secure under these standard intractability assumptions. Refer to [MW99, Bon98] for further reading.

Now suppose that two distinct cyclic groups $\mathcal{G}_1 = \langle g_1 \rangle$ and $\mathcal{G}_2 = \langle g_2 \rangle$ of prime order q are a bilinear group pair. We then obtain natural generalizations of the computational and decisional Diffie-Hellman problems with this bilinear group pair $(\mathcal{G}_1, \mathcal{G}_2)$. Using the prefix "co" implies that the given two groups are different (but with the same order).

Computational Co-Diffie-Hellman Problem (co-CDH). For $a \in \mathbb{Z}_q$, given $g_1, g_1^a \in \mathcal{G}_1$ and $h \in \mathcal{G}_2$, compute $h^a \in \mathcal{G}_2$.

Decision Co-Diffie-Hellman Problem (co-DDH). For $a, b \in \mathbb{Z}_q$, given g_1, $g_1^a \in \mathcal{G}_1$ and $h, h^b \in \mathcal{G}_2$, answer with;
   (i) Yes, if $a = b$. In this case, (g_1, g_1^a, h, h^b) is called a co-Diffie-Hellman quadruple.
   (ii) No, otherwise.

Here, we define co-gap Diffie-Hellman (co-GDH) groups to be group pairs $\mathcal{G}_1$ and $\mathcal{G}_2$ on which co-DDH is easy but co-CDH is hard. This class of problems— gap problems—was first introduced by Okamoto and Pointcheval [OP01] and is considered to be dual to the class of the decision problems. They also discussed several cryptographic applications based on these problems such as the undeniable signature scheme introduced by Chaum and Pederson [CP93] and designated confirmer signatures. Joux and Nguyen [JN01] also pointed out that, for the special case of supersingular elliptic curves, DDH is easy while CDH is still hard.

Definition 2 (co-GDH assumption [BGLS03]). *We say that both $\mathcal{G}_1$ and $\mathcal{G}_2$ are (t, ϵ)-co-GDH groups if they are in the class of decision groups for co-Diffie-Hellman and no PPT algorithm (t, ϵ)-breaks co-CDH in groups $\mathcal{G}_1$ and $\mathcal{G}_2$ with sufficiently large prime q.*

2.3 Co-GDH Signature Scheme

The co-GDH signature scheme due to Boneh *et al.* [BLS04] is known to be one of the shortest signatures in the literature. In this scheme, there are three algorithms, *KeyGen*, *Sign*, and *Verify*, plus a full-domain hash function $H : \{0,1\}^* \to \mathcal{G}_2$, which works as a random oracle.

Key Generation. The secret key x is chosen at random $x \in_R \mathbb{Z}_q$, and the corresponding public key y is computed by $y = g_1^x$. The public key is an element of $\mathcal{G}_1$.

Signing. To generate a signature on a message $M \in \{0,1\}^*$ under the secret key x, compute $\sigma = h^x$, where $h = H(M)$. Thus, the message-signature pair becomes (M, σ).

Verification. The verification algorithm takes the signer's public key y, message M, and signature σ as input. It first computes $h = H(M)$ and checks if (g_1, y, h, σ) is a valid co-Diffie-Hellman quadruple.

We can simply make use of a bilinear map $\hat{e}$ from $\mathcal{G}_1 \times \mathcal{G}_2$ to $\mathcal{G}_T$ for co-DDH testing: $\hat{e}(g_1, \sigma) \overset{?}{=} \hat{e}(y, h)$. This signature scheme is existentially unforgeable under adaptive chosen message attacks [GMR88] in the random oracle model. However, the security reduction is loosely related to the CDH problem.

2.4 Proving Equality of Two Discrete Logarithms

Let $\mathcal{G} = \langle g \rangle = \langle h \rangle$ be a group of prime order q with generators g, h. Denote $\mathsf{EDL}(g, h)$ as the language of pairs $(y, \sigma) \in \mathcal{G}^2$ such that $\mathtt{dlog}_g y = \mathtt{dlog}_h \sigma$, where $\mathtt{dlog}_g y$ denotes the discrete logarithm of y with respect to the base g (identically for $\mathtt{dlog}_h \sigma$). The following is a well-known zero-knowledge proof system for the proof of equality of two discrete logarithms [CP93]. This system proceeds as follows.

- Given $(y, \sigma) \in \mathsf{EDL}(g, h)$, the prover picks $k \in_R \mathbb{Z}_q$ at random, computes $u = g^k, v = h^k$, and then sends them to the verifier.
- The verifier chooses $c \in_R \mathbb{Z}_q$ as a public random challenge and sends it to the prover.

- Upon receipt of c, the prover responds with $s = k + cx$.
- The verifier checks if $g^s \stackrel{?}{=} uy^c$ and $h^s \stackrel{?}{=} v\sigma^c$ hold true.

It is well known that this proof system is both complete and sound. See [CP93] for details. Notice that this interactive proof system is a public-coin zero-knowledge since a simulator given inputs g, h, y, σ can choose c, s at random in $\mathbb{Z}_q$ and compute u and v from them. A non-interactive version of this proof system has several applications such as EDL signature schemes [GJ03, KW03, CM05] that are proved to be as secure as the CDH problem and threshold cryptosystems that are provably secure against chosen ciphertext attacks under the CDH and DDH assumptions [SG98], viewing the hash functions as random oracles [BR93].

3   P2DL Signature Scheme

3.1   New Model

The participants in our protocol are classified into three groups; *Signer*, *Verifier*, and *Notary*. Each of them plays a respective role:

Signer. She wants to produce a signature on a document in an ad-hoc network. Each signer has its ad-hoc and infra certificates that are issued by the respective ad-hoc and infra CAs. Roughly speaking, her P2DL signature contains three essential ingredients:
  1. An ad-hoc signature σ_1. This is only valid in the ad-hoc network.
  2. An infra signature σ_2. This is only valid in the infra network. She does not want any recipient to obtain this signature until notarization. Thus, an encrypted signature is transmitted to a receiver.
  3. A zero-knowledge argument is provided to prove herself possessing a couple of signing (secret) keys, which are used in generating σ_1 and σ_2 signatures.
  In general, she does not have to establish a communication channel to the infra network.
Verifier. Any verifier is assumed to be potentially mobile from the ad-hoc network to the infra network. Upon receipt of a signer's signature, the verifier can immediately verify her ad-hoc signature σ_1 within an ad-hoc network. After switching to the infra network, the verifier can obtain an infra signature σ_2 by asking the trusted notary to translate P2DL signature into an infra one.
Trusted Notary. A special-purpose TTP is assumed, the trusted notary.[1] The notary operates when a verifier makes a request of notarization. The notary performs two tasks:

---

[1] Notice that the notary is an entity only available in the infra network while signers stay within the ad-hoc network (when sending their signatures to verifiers). In contrast, verifiers are possibly mobile from an ad-hoc network to the infra network. Our signature scheme features the mobility nature of verifiers.

1. It determines whether the given P2DL signature is valid and whether the given ad-hoc and infra signatures are generated by one person. If "YES," it publishes σ_2 as the translation of σ_1. If "NO," it notifies that the given signature is invalid.
2. During notarization, the notary offers a public-key certification service upon receiving the signer's (infra) public key on behalf of the on-line (infra) CA.

3.2 P2DL Signature Scheme

The protocol uses two hash functions. The security analysis will view these hash functions as random oracles.

$$H_1 : \{0,1\}^* \times \mathbb{Z}_q \to \mathcal{G}_2, \quad H_2 : (\mathcal{G}_2)^9 \to \mathbb{Z}_q$$

Key Generation. Each signer is assumed to hold two (distinct) key pairs, which are used in an ad-hoc network and in the infra network separately. For a particular signer, pick $x_2 \in_R \mathbb{Z}_q$ at random, and compute $y_2 = g_1^{x_2}$. The infra secret key is $x_2 \in \mathbb{Z}_q$ and the corresponding public key is $y_2 \in \mathcal{G}_1$. Then, the ad-hoc key and trusted notary's key pairs, (x_1, y_1) and (x_N, y_N), are generated, respectively, in the same manner as the infra key generation. Note that an ad-hoc key might be linearly independent of the infra secret.

Signature Generation. Given the key pairs (x_1, y_1) and (x_2, y_2), and the trusted notary's public key y_N, the algorithm to sign a message $M \in \{0,1\}^*$ runs as follows. The signer chooses $r_1, r_2, k_1, k_2 \in_R \mathbb{Z}_q$ at random, and computes

$$h = H_1(M, r_1), \ \sigma_1 = h^{x_1}, \ \sigma_2 = h^{x_2}, \ u_1 = g_2^{k_1}, \ v_1 = h^{k_1}, \ u_2 = g_2^{k_2}, \ v_2 = h^{k_2},$$
$$c = H_2(h, \psi(y_1), \psi(y_2), \sigma_1, \sigma_2, u_1, v_1, u_2, v_2), \ s_1 = k_1 + cx_1, \ s_2 = k_2 + cx_2,$$
$$w = \langle \ g_2^{r_2}, \sigma_2 \cdot \psi(y_N)^{r_2} \ \rangle.$$

The P2DL signature is $\Sigma = (r_1, \sigma_1, w, s_1, s_2, c)$. This will be sent to the verifier along with the message M and *implicitly* public keys y_1, y_2, and y_N.

At first glance, we can see two individual co-GDH signatures σ_1, σ_2 generated under the ad-hoc and infra secrets, respectively. However, the signer does not want any recipient to "instantly" possess her infra signature σ_2 until the trusted notary permits publication of σ_2. Thus, σ_2 is encrypted into w with the notary's public key, while σ_1 is not. The ciphertext w is exactly in the form of an ElGamal type ciphertext [ElG85]. In addition, a non-interactive zero-knowledge argument proving that $(y_1, \sigma_1) \in \mathsf{EDL}(g_1, h)$ and, simultaneously, $(y_2, \sigma_2) \in \mathsf{EDL}(g_1, h)$ is presented to demonstrate the signer's possession of two discrete logarithms x_1, x_2. Indeed, this ZK argument works as a nexus or correlation between σ_1 and σ_2.

Verification. Upon receipt of (M, Σ), any user can verify σ_1. First, a verifier parses $\Sigma \xrightarrow{\mathrm{P}} r_1 \| \sigma_1 \| w \| s_1 \| s_2 \| c$. The verifier then computes $h = H_1(M, r_1)$ and checks if

$$\hat{e}(g_1, \sigma_1) \stackrel{?}{=} \hat{e}(y_1, h). \tag{1}$$

Consistency is easily proved because $\hat{e}(g_1, \sigma_1) = \hat{e}(g_1, h)^{x_1} = \hat{e}(g_1^{x_1}, h) = \hat{e}(y_1, h)$. However, the infra signature σ_2 cannot be verified instantly because it is encrypted. One can verify the *encrypted* signature instead since the encrypted infra signature in our scheme is analogous to the bilinear verifiably-encrypted signature.[2] Perform the following: parse $w \to U\|V$; check if the equation below holds.

$$\hat{e}(g_1, V) \overset{?}{=} \hat{e}(y_2, h) \cdot \hat{e}(y_N, U) \tag{2}$$

This also ensures consistency of verification for any valid encrypted signature; $\hat{e}(g_1, V) = \hat{e}(g_1, \sigma_2 \cdot \psi(y_N)^{r_2}) = \hat{e}(g_1, h^{x_2}) \cdot \hat{e}(g_1, \psi(y_N)^{r_2}) = \hat{e}(y_2, h) \cdot \hat{e}(g_1, U^{x_N}) = \hat{e}(y_2, h) \cdot \hat{e}(y_N, U)$.

The verifier, voluntarily, forwards (M, Σ) to the trusted notary when he or she wishes to confirm the authorship of the ad-hoc signature σ_1 and to obtain a "translation" of σ_1 available in the infra network. The notary in turn responds with the validity of the signature and the infra version of the given signature i.e., σ_2 as required. We refer to this algorithm as *Notarization*.

Notarization. The notarization algorithm takes as input the notary's decryption key x_N, the signer's ad-hoc and infra public keys y_1, y_2, and (M, Σ). After parsing the P2DL signature, the notary performs the following:

- Check whether the public key y_2 is revoked (e.g, OCSP [MAM+99] and CRL profile [HFPS99]); If so, output $\perp$ and terminate;
- Otherwise, parse the ciphertext into two components such that $w \overset{\text{P}}{\to} U\|V$;
- Decrypt the ciphertext: $\sigma_2 = V/U^{x_N}$;
- Compute $h = H_1(M, r_1)$;
- Compute $u_1 = g_2^{s_1} \psi(y_1)^{-c}$, $v_1 = h^{s_1} \sigma_1^{-c}$, $u_2 = g_2^{s_2} \psi(y_2)^{-c}$, $v_2 = h^{s_2} \sigma_2^{-c}$;
- Compute $c' = H_2(h, \psi(y_1), \psi(y_2), \sigma_1, \sigma_2, u_1, v_1, u_2, v_2)$;
- If $c' = c$, publish σ_2 as the infra signature; Otherwise, output $\perp$.

This algorithm runs when the verifier has made a request to the trusted notary for the aforementioned reasons. The notarization algorithm consists of two types of checking procedures. First, it checks whether the signer's infra public key is revoked. Second, the notary checks if the proof system assures that $(y_1, \sigma_1) \in \mathsf{EDL}(g_1, h) \wedge (y_2, \sigma_2) \in \mathsf{EDL}(g_1, h)$. Equivalently, if the zero-knowledge is proved, it convinces the notary that the signer owns both $\mathtt{dlog}_{g_1} y_1$ and $\mathtt{dlog}_{g_1} y_2$. If all the checking procedures output "YES," the notary determines to publish σ_2 as the rendering of σ_1 on behalf of the signer. Alternatively, the trusted notary appends its signature on this infra signature σ_2 to enable every user to distinguish the notary's publications from ordinary (infra) signatures.

---

[2] The bilinear verifiably-encrypted signature [BGLS03] is a simple combination of the co-GDH signature and ElGamal cryptosystem. Our design goal is to provide the universal verifiability of digital signatures across two domains, while the verifiably-encrypted signature is applicable to the contract signing protocol with one domain setting. Moreover, the resulting signature scheme features a *tight* security reduction unlike [BGLS03].

Confirmation. Upon satisfaction of notarization, the verifier receives an infra signature σ_2 from the notary. The verifier checks if the following equation holds true

$$\hat{e}(g_1, \sigma_2) \stackrel{?}{=} \hat{e}(y_2, h). \tag{3}$$

The signature is verified under the signer's public key, which has already been certified. The consistency in verifying σ_2 is easily provable as we have likewise done for (1).

4 Security Analysis

4.1 Proving Possession of Two Discrete Logarithms

The zero-knowledge proof of the equality of two discrete logarithms introduced in Sect 2.4 helps a prover to convince a verifier of his possession of a *single* secret exponent $x \in \mathbb{Z}_q$ that is the secret key of the corresponding public key y and the discrete logarithm of another element σ with respect to the base h as well: $x = \mathrm{dlog}_g y = \mathrm{dlog}_h \sigma$. The P2DL signature scheme, however, requires a non-interactive zero-knowledge proof of the possession of two (not single but distinct) discrete logarithms x_1 and x_2, which is an extension of the zero-knowledge proof of the equality of two discrete logarithms. We refer to this special honest-verifier proof system as P2DL.

The P2DL proof system takes as input the prover's secret exponents x_1 and x_2 as elements of a certain group $\mathcal{G}$, two generators g and h of $\mathcal{G}$, a hash function H_2 that maps from $(\mathcal{G})^9$ to $\mathbb{Z}_q$. First, the prover picks $k_1, k_2 \in_R \mathbb{Z}_q$ at random. Next, proceed the following in order to prove that $(y_1, \sigma_1) \in \mathsf{EDL}(g, h) \wedge (y_2, \sigma_2) \in \mathsf{EDL}(g, h)$:

$$u_1 = g^{k_1}, \ v_1 = h^{k_1}, \ u_2 = g^{k_2}, \ v_2 = h^{k_2},$$
$$c = H_2(h, y_1, y_2, \sigma_1, \sigma_2, u_1, v_1, u_2, v_2),$$
$$s_1 = k_1 + cx_1, \ s_2 = k_2 + cx_2.$$

The binary operator "$\wedge$" implies the logical operator "AND." The resulting proof is (c, s_1, s_2) and can be verified by first reconstructing the commitments

$$u_1' = g^{s_1} y_1^{-c}, \ v_1' = h^{s_1} \sigma_1^{-c}, \ u_2' = g^{s_2} y_2^{-c}, \ v_2' = h^{s_2} \sigma_2^{-c},$$

where $y_1 = g^{x_1}$ and $y_2 = g^{x_2}$, and then checking the equations below

$$c' = H_2(h, y_1, y_2, \sigma_1, \sigma_2, u_1', v_1', u_2', v_2') \quad \text{and} \quad c' \stackrel{?}{=} c.$$

The verifier accepts if $c' = c$; otherwise, the verifier rejects.

Theorem 1. *The P2DL ZK protocol mentioned above is a special honest-verifier proof system for proving that $(y_1, \sigma_1) \in EDL(g, h) \wedge (y_2, \sigma_2) \in EDL(g, h)$, assuming the couple of keys belongs to a user.*

Proof. Completeness of the protocol can be easily seen so that an honest prover will always succeed in constructing a valid proof since

$$u_1' = g^{s_1}y_1^{-c} = g^{k_1+cx_1}g^{-cx_1} = g^{k_1} = u_1,$$
$$v_1' = h^{s_1}\sigma_1^{-c} = h^{k_1+cx_1}h^{-cx_1} = h^{k_1} = v_1.$$

Likewise, we can see the correctness for u_2' and v_2'. Therefore, for an honest prover, $c' = c$. Let us now consider the soundness of P2DL ZK protocol. Suppose that a cheating prover who knows neither x_1 nor x_2 can generate another correct response and challenge pair $(\hat{c}, s_1', s_2')$ that differs from the given (c, s_1, s_2), then

$$\begin{cases} g^{s_1-s_1'} = y_1^{\hat{c}-c}, \ g^{s_2-s_2'} = y_2^{\hat{c}-c}, \\ h^{s_1-s_1'} = \sigma_1^{\hat{c}-c}, \ h^{s_2-s_2'} = \sigma_2^{\hat{c}-c}, \end{cases}$$

and hence

$$\begin{cases} \mathtt{dlog}_g y_1 = \mathtt{dlog}_h \sigma_1 = \frac{s_1-s_1'}{\hat{c}-c}, \\ \mathtt{dlog}_g y_2 = \mathtt{dlog}_h \sigma_2 = \frac{s_2-s_2'}{\hat{c}-c}. \end{cases}$$

This contradicts the assumption that the cheating prover knows neither x_1 nor x_2. Thus, the soundness probability is at most $1/q^2$. Even when a prover knows one of the discrete logarithms, the prover's probability of successfully cheating is at most approximately $1/q$. $\qquad\square$

In the actual run of P2DL signature scheme, a slightly modified version of this proof system is used, where an isomorphism ψ and two distinct groups $\mathcal{G}_1 = \langle g_1 \rangle$ and $\mathcal{G}_2 = \langle g_2 \rangle$ with a prime order q might be used.

4.2   Security Against Existential Forgeries

Boneh *et al.* have shown that the co-GDH signature scheme is secure, namely existentially unforgeable, under a chosen message attack as long as the co-GDH assumption holds. The P2DL signature introduced here uses the co-GDH signature as the underlying signature scheme. However, it still remains unproven that an attacker who has at most one of the secret keys x_1 and x_2 cannot generate any valid P2DL signature tuple $\Sigma = (r_1, \sigma_1, w, s_1, s_2, c)$ with a non-negligible probability.

Theorem 2. *The P2DL signature scheme is provably secure against existential forgery under adaptive chosen message attacks assuming that (a) all the hash functions are chosen from the random oracles, and (b) the co-GDH assumption holds.*

The fully detailed proof of this theorem is described in the full version of this paper [LOM05].

5 Concluding Remarks

Coalition-Resistance. Observe that a colluding couple of ad-hoc and infra users can generate a valid (but forged) P2DL signature since the couple of used key pairs for ad-hoc and infra users is completely *loosely-coupled.* Therefore, we should make any relationship between the signing keys. A simple countermeasure we have in mind is to provide an additional administration policy on issuing ad-hoc certificates. Any signer A has the following formatted certificate, which is denoted by

$$\text{CERT}_A(\text{ad-hoc}) = \langle ID_1, PK_1, \text{CERT}_A(\text{infra}), \text{params}, Sig_{\text{CA}_1}(\text{all parts thereof})\rangle$$

where $ID_1 = ID_A(\text{ad-hoc})$ and $PK_1 = PK_A(\text{ad-hoc})$. In addition, params represents the domain parameters $(\mathcal{G}_1, \mathcal{G}_2, \mathcal{G}_T, g_1, g_2, \psi, \hat{e}, q)$ and $Sig_{\text{CA}_1}(\cdot)$ denotes an (ordinary) signature of the ad-hoc CA. Whenever issuing an ad-hoc certificate, it is mandatory for the ad-hoc CA to authenticate whether the user is the owner of the predetermined infra certificate

$$\text{CERT}_A(\text{infra}) = \langle ID_2, PK_2, \text{params}, Sig_{\text{CA}_2}(\text{all parts thereof})\rangle.$$

Each of ID_2, PK_2, and $Sig_{\text{CA}_2}(\cdot)$ has the equivalent meaning to ID_1, PK_1, and $Sig_{\text{CA}_1}(\cdot)$ in the infra network. Note that the same domain parameters params are used even in the infra network. Achieving the coalition-resistance via more sophisticated constructions such as group signatures [ACJT00, BBS04] can be a compelling approach, but it is not of our interest.

Towards Heterogeneous Settings. Our protocol is constructed only within a "homogeneous" mathematical setting. In other words, both in ad-hoc network and in infra network, the same domain parameters and mathematical functions are appropriately assumed. Nonetheless, it would be more preferable to consider the universal verifiability among "heterogeneous" domains with distinct domain parameters to achieve the so-called *ubiquitous* security. As far as we know, this generalized construction still remains unanswered.

References

[ACJT00]   Giuseppe Ateniese, Jan Camenisch, Marc Joye, and Gene Tsudik. A practical and provably secure coalition-resistant group signature scheme. In *Advances in Cryptology – CRYPTO '00*, volume 1880 of *Lecture Notes in Computer Science*, pages 255–270. Springer-Verlag, August 2000.

[BBS04]   Dan Boneh, Xavier Boyen, and Hovav Shacham. Short group signatures. In *Advances in Cryptology – CRYPTO '04*, volume 3152 of *Lecture Notes in Computer Science*, pages 41–55. Springer-Verlag, August 2004.

[BGLS02]   Dan Boneh, Craig Gentry, Ben Lynn, and Hovav Shacham. Aggregate and verifiably encrypted signatures from bilinear maps. Cryptology ePrint Archive, Report 2002/175, 2002. http://eprint.iacr.org/.

[BGLS03]   Dan Boneh, Craig Gentry, Ben Lynn, and Hovav Shacham. Aggregate and verifiably encrypted signatures from bilinear maps. In *Advances in Cryptology – EUROCRYPT '03*, volume 2656 of *Lecture Notes in Computer Science*, pages 416–432. Springer-Verlag, May 2003.

[BLS04]   Dan Boneh, Ben Lynn, and Hovav Shacham. Short signatures from the Weil pairing. *Journal of Cryptology*, 14(4):297–319, 2004.

[Bon98]   Dan Boneh. The decision Diffie-Hellman problem. In *Proceedings of the Third Algorithmic Number Theory Symposium*, volume 1423 of *Lecture Notes in Computer Science*, pages 48–63. Springer-Verlag, 1998.

[BR93]   Mihir Bellare and Phillip Rogaway. Random oracles are practical: A paradigm for designing efficient protocols. In *Proceedings of ACM Conference on Computer and Communications Security '93*, pages 62–73. ACM Press, 1993.

[CM05]   Benoît Chevallier-Mames. An efficient CDH-based signature scheme with a tight security reduction. In *Advances in Cryptology – CRYPTO '05*, volume 3621 of *Lecture Notes in Computer Science*, pages 511–526. Springer-Verlag, August 2005.

[CP93]   David Chaum and Torben Pryds Pedersen. Wallet databases with observers. In *Advances in Cryptology – CRYPTO '92*, volume 740 of *Lecture Notes in Computer Science*, pages 89–105. Springer-Verlag, August 1993.

[DH76]   Whitfield Diffie and Martin Hellman. New directions in cryptography. *IEEE Transactions on Information Theory*, 22(6):644–654, November 1976.

[ElG85]   Taher ElGamal. A public key cryptosystem and a signature scheme based on discrete logarithms. *IEEE Transactions on Information Theory*, 31(4):469–472, July 1985.

[GJ03]   Eu-Jin Goh and Stanisław Jareki. A signature scheme as secure as the Diffie-Hellman problem. In *Advances in Cryptology – EUROCRYPT '03*, volume 2656 of *Lecture Notes in Computer Science*, pages 401–415. Springer-Verlag, May 2003.

[GMR88]   Shafi Goldwasser, Silvio Micali, and Ronald Rivest. A digital signature scheme secure against adaptive chosen-message attacks. *SIAM Journal on Computing*, 17(2):281–308, 1988.

[HFPS99]   R. Housley, W. Ford, W. Polk, and D. Solo. Certificate and CRL profile. RFC 2459, January 1999. `http://www.ietf.org/`.

[ITU97]   X.509 (1997 e): Information Technology – Open Systems Interconnection – The Directory: Authentication Framework, 1997.

[JN01]   Antoine Joux and Kim Nguyen. Separating decision Diffie-Hellman from Diffie-Hellman in cryptographic groups. Cryptology ePrint Archive, Report 2001/003, 2001. `http://eprint.iacr.org/`.

[KW03]   Jonathan Katz and Nan Wang. Efficiency improvements for signature schemes with tight security reductions. In *Proceedings of ACM Conference on Computer and Communications Security '03*, pages 155–164. ACM Press, 2003.

[LOM05]   KyungKeun Lee, JoongHyo Oh, and SangJae Moon. How to generate universally verifiable signatures in ad-hoc networks. Cryptology ePrint Archive, Report 2005/389, 2005. `http://eprint.iacr.org/`.

[MAM+99]   M. Myers, R. Ankney, A. Malpani, S. Galperin, and C. Adams. Online certificate status protocol - OCSP. RFC 2560, 1999. `http://www.ietf.org/`.

[MW99]   Ueli Maurer and Stefan Wolf. The relationship between breaking the
            Diffie-Hellman protocol and computing discrete logarithms. *SIAM Jour-
            nal on Computing*, 28(5):1689–1721, 1999.
[OP01]   Tatsuaki Okamoto and David Pointcheval. The gap-problems: a new
            class of problems for the security of cryptographic schemes. In *In-
            ternational Workshop on Practice and Theory in Public Key Cryptog-
            raphy – PKC'01*, volume 1992 of *Lecture Notes in Computer Science*,
            pages 104–118. Springer-Verlag, February 2001.
[SG98]   Victor Shoup and Rosario Gennaro. Securing threshold cryptosystems
            against chosen ciphertext attack. In *Advances in Cryptology – EU-
            ROCRYPT '98*, volume 1403 of *Lecture Notes in Computer Science*,
            pages 1–16. Springer-Verlag, May 1998.
[ZH99]   Lidong Zhou and Zygmunt J. Haas. Securing ad hoc networks. *IEEE
            Network*, 13(6):24–30, November-December 1999.

"Fair" Authentication in Pervasive Computing

Jun Li[1], Bruce Christianson[1], and Martin Loomes[2]

[1] School of Computer Science,
University of Hertfordshire,
Hatfield, Herts,
United Kingdom AL10 9AB
{j.7.li, b.christianson}@herts.ac.uk
[2] School of Computing Science,
Middlesex University,
The Burroughs, London,
United Kingdom NW4 4BT
m.loomes@mdx.ac.uk

Abstract. Authentication is traditionally required to be strong enough to distinguish legitimate entities from unauthorised entities, and always involves some form of proof of identity, directly or indirectly. Conventional storable or delegable authentication scenarios in the pervasive computing environment are often frustrated by the qualitative changes of pervasive computing when humans are admitted into the loop. In this paper, we present an alternative approach based upon involving *human self-determination* in security protocols. This targets the authentication problem in pervasive computing, particularly when communication occurs in mobile ad-hoc fashion. We propose the argument of "thinkable" authentication, which involves using two-level protocols with the consideration of minimising trustworthiness in both human and computer device domains, but without unnecessary entity identity authentication. Thus, self-determining knowledge of the human interactions in pervasive computing can be exploited in order to make improvements on current security mechanisms.

Keywords: Authentication, Pervasive Computing, Mobile ad-hoc Networks, Trust, Human self-determination.

1 Introduction

Weiser's vision [31] of computer devices equipped with wireless networking capability seamlessly pervading society and people's lifestyles, is in the process of becoming reality. Various, and numerous, roaming computer devices surrounding humans, as Digital Representative Devices (DRDs)[1], will contribute qualitative differences to convenient, integrated and invisible communication.

---

[1] The term "DRD" in this paper simply refers to an arbitrary computer device with wireless communications capability. Such devices range from current mobile phones, PDAs, and laptops to future advanced devices.

M. Burmester and A. Yasinsac (Eds.): MADNES 2005, LNCS 4074, pp. 132–143, 2006.

In such a pervasive computing[2] environment, dynamic "trustworthiness" relationships between humans and humans, devices and devices, and humans and devices, are increasingly important when convincing people to adopt new technologies. "Trustworthiness" is a quite sensitive word in the security field, having different meanings in different contexts (our understanding will be given in section 5). In this paper, we argue that adopting the conventional assumptions of independently computed trustworthiness for DRDs would actually undermine security performance and deceive security protocol design. More significantly, we propose and assess an alternative approach, based upon involving *human self-determination* in security protocols, targeted to authentication in pervasive computing.

Michael Roe in his thesis [23] shows that what is regarded as a security threat in one context may become a mechanism providing a desired security service in a different context. Despite the fact that human influences are usually considered as the "threats", our concept of "human self-determination" is an attempt to positively re-frame knowledgable human interactions as a desirable security mechanism in the context of mobile ad-hoc network protocols, in a similar manner to Roe's "threat/service duality".

Authentication is currently a fundamental building block used to support other security properties, such as confidentiality, integrity and availability [25]. Authentication was originally introduced to guarantee the communicating parties are who they claim to be, implicitly or explicitly, in a relatively static environment [22,6,18]. Current authentication protocols[3] involve some form of proof of identity, directly or indirectly, in most cases. In traditional computing, which largely ignores the semantics of the human interactions that it supports, authentication is always required to be strong enough to distinguish legitimate entities from unauthorised entities, mostly by relying on challenge-response identification or interactive proofs involving Trusted Third Parties (TTPs). However, strong authentication is an inefficient, heavyweight task with strict requirements for use of a progression of computation resources and standardised infrastructures, which are increasingly frustrating to fulfil in the pervasive computing environment.

The features of pervasive computing, (such as decentralised ad-hoc architectures, dynamic enrolments, transient associations, unreliable wireless access paths, limited device resources), will entail a massive qualitative change in addition to quantitative change [11,15,24,27]. It is very difficult from an engineering perspective to accomplish traditional authentication scenarios in this environment. Among the more serious reasons which make cryptosystems fail in the real world are implementation errors and management failures [1]. This situation becomes worse as roaming DRDs increasingly pervade human interaction. Because of the uncertainty of human behaviour, the computed trustworthiness of devices has been excessively emphasised in traditional computing, omitting

---

[2] We use "pervasive computing" in this paper as a synonym for ubiquitous computing or ubicom.

[3] For this paper's purpose, the term "authentication" refers to mutual entity authentication (rather than e.g. data authentication) if no specific indication is given.

consideration of human involvement. Current wide-spread authentication mechanisms have unavoidable weaknesses which make them unsuitable to be deployed in pervasive computing. We have to conceive a new paradigm to implement authentication procedures. In our view, the "threat" of human influences can also be considered as a security service (in a similar manner to the "threat/service duality" stated in [23]). Thus, self-determining knowledge of the human interactions in pervasive computing can be exploited in order to make improvements on security mechanisms.

This paper starts with a brief review of previous approaches. In section 3 we give an overview of storable and delegable authentication scenarios in traditional computing, followed by a description of the paradigm shift of authentication in the pervasive computing environment in Section 4. The consideration of minimising trustworthiness will be addressed in section 5. We explain our argument of "fair" authentication, unnecessary entity identity authentication and "thinkable" authentication, with the support of two-level protocols in Section 6. Finally, future work and conclusions are set out in section 7.

2   Previous Approaches

Novel approaches to authentication in pervasive computing, e.g. in mobile ad-hoc environments, has been an interesting issue and addressed in many papers in different ways. For instance, Balfanz *et al.* [4] proposed the use of location-limited channels, and pre-authentication which is based on "Resurrecting Duckling" [27], aiming to bootstrap trust between strangers. In their approach, devices exchange some public information over a location-limited side channel, then make use of such information to implement authentication protocols with public key cryptography over the wireless communication channel. Meanwhile, twin-channel threat models[4] in ubiquitous computing, where one is the communication medium channel with unreliable security and the other is a more costly channel with higher security, was suggested in [10] with the sampled protocols involving public key certificates. Other related work is found in the scenario of bootstrapping with a shared weak-secret (e.g. password) but achieving strong authentication between two devices [9,3,2,13]. These key agreement protocols provide a good way to consider authentication in the circumstances which do not involve third parties. Nevertheless, most previous work still involves the negative impact of identity authentication, and none of them specify and formalise the significance of the role of human self-determination for authentication protocols in pervasive computing.

3   Authentication in Traditional Computing

A list of conventional authentication techniques in traditional computing is given in paper [11], which includes shared secrets, public key cryptography schemes, tokeni-

---

[4] Wong and Stajano also introduced multi-channel protocols to achieve strong authentication between two wireless devices (at least one device is camera-equipped) in the 2005 Security Protocols Workshop in Cambridge.

sation and biometrics. Here, we categorise these considerably developed authentication techniques into two scenarios, *Storable scenario* and *Delegable scenario*.

3.1   Storable Authentication

The storable type of authentication is straightforward and easy to use, not requiring further costs for users. It includes password, biometrics, and secret key systems such as Kerberos[5] [19]. Basically in these approaches, it is necessary to keep a centralised database of information based upon users' secrets (e.g. hash value of passwords, users' secret keys) and of personal biometric data. The decision of authentication is based on the results of required input information computationally matching pre-stored information in the centralised database. Thus, strong security assumptions for both challenge-response identification communication paths and remote servers are required. The problem of the storable scenario, however, is the *Confinement* problem, since the compromise of one assumption (for example by penetration of a communication path) will bring the collapse of trust in the entire architecture.

3.2   Delegable Authentication

Compared with the storable scenario, delegable authentication has earned much wider application due to its strong authentication performance with the assistance of public key cryptography. It includes public key certificates (as applied in PKIs and PGP), and tokenisation (or smart cards). Furthermore, we also categorise capability-based systems [21,16,14] into this scenario, due to their crucial relation between authentication and access control. Capability systems were originally oriented towards authorisation rather than authentication, to the point that many papers do not regard them as authentication mechanisms at all. We shall return to this point below. For the delegable scenario, the output of the authentication process is justified by the input of authorised delegated capabilities being recognised. It scales the proof of capability (e.g. certificate chain) from the dedicated Trusted Third Parties (TTPs) or authorities in a transitive way. The "triangle" of trust among prover, verifier and global TTPs in one domain, is usually required during the authentication process. The problem of the delegable scenario is *Trust Transitivity*, which results in implementation difficulty (e.g. global TTPs topology) and uncontrolled imposition of trust (e.g. the unfair compulsion to fully trust arbitrary TTPs), particularly with the communication crossing different trust domains.

4   Paradigm Shift in Pervasive Computing

4.1   Dynamic Trust Binding

A number of costly scenarios and systems have been designated to authenticate communicating entities' identities, either universal names or logical identities.

---

[5] In terms of the "ticket" element, Kerberos can also be considered under the second scenario.

These aim to secure the association between legitimate users' identities and their resources. This kind of association attempts to secure the entire communication from very beginning but is always subject to threats, such as ID-theft and spoofing. Generally, Bob masquerades as Alice not because Bob really fancies being Alice. Instead, Bob is interested in the resources or access rights associated with Alice's identity.

A much better binding of "people → key → capability" has been described in [8] and is usually deployed in the delegable scenario (e.g. SPKI as seen in [12]). However, the delegable scenario has suffered increasingly from vulnerability to classic "man-in-the-middle" attacks, since a "mailman" can manually construct a delegation binding as, "people → key → key2 → capability", where "key2" is actually controlled by the mailman. Communicating parties and protocols will find it increasingly difficult to notice this difference particularly when a user interface within a device behaves as a hidden mailman. Intuitively, people have worried that lost or stolen DRDs will give masqueraders opportunities to breach their privacy and abuse their privileges. Binding public keys with identity within public key certificates is an effective and efficient way to authenticate entity in traditional computing. Consequently, many researchers have tried to modify traditional certificate-based schemes in order to meet the requirements of pervasive computing from the engineering perspective. But this demands an accessible path to TTPs whensoever the protocols need, not just to obtain the corresponding entities' certificates, but also to cope with more serious revocation circumstances. Furthermore, a common requirement of pervasive computing is to provide flexible communication with "strangers" in dynamically changing wireless ad hoc fashion. More likely, communicating entities are coming from different trust domains, holding a variety of certificates issued by their own TTPs. Even if the recognition of certificates crossing domains is solved by chained negotiation among TTPs, there is no guarantee that the success of security policy checking in one domain will be propagated across other trust domains (see section 5 below). Thus, in addition to the reliable communication paths, the unfair Trust Transitivity problem [5,7] indicates that the pre-issued certificate-like delegable authentication approaches have their own weaknesses for dynamic trust binding in pervasive computing.

4.2   Human-Invisibility vs. Computer-Invisibility

In some sense, humans are invisible in traditional computing contexts such as Internet-based computing. Computer devices follow human instructions, but ignore whether instructions are appropriate to the tasks or come from the right human. Thus, strong identity-based authentication is always required to ensure that a tracing step can be followed if something is going wrong. In wireless ad hoc networks (as a part of pervasive computing), it is very clear that users always have pre-decision ("This is the one to whom I am willing to talk") and physico-spatial knowledge ("Yes, I can see this is the one I am going to talk to"). Nowadays, pervasive computing is characterised with the achievement of computer-invisibility, people communicating with the presence of physical visible DRDs but without noticing their existence. In other words, it is desirable to transform security techniques into the new human-based philosophy for pervasive computing.

5   "Maltrust" or Minimise Trustworthiness

Superficially, limiting human influences on computing systems is usually a basic discipline to guide security protocol design, especially for authentication protocols in traditional computing. It is considered reasonable because of human unpredictability, including dishonest or incompetent behaviours. A "maltrust problem" occurs when humans abuse "trust" gained from other humans. Most traditional computing schemes are built upon computed credentials from computing devices, intending to solve the "maltrust problem". A typical example in the real world is the current Chip and PIN credit card approach, shifting jurisdiction from human verification (signature recognition) to computed authentication (system verifying PIN matching). These schemes, however, have not achieved better security performance because the essential "maltrust problem" has not been solved as it was expected to be, but is simply reproduced from the humans-humans domain to the humans-devices domain. Consequently, increasing human reliance upon computer devices with the seamless interactions between humans and DRDs in pervasive computing is in fact compounding the problems caused by maltrust.

The principle of our proposal is based on a "Need-to-Know" policy[6]. This is not new, and was originally produced in a military context and classically applied in access control systems via minimising access rights. Note that in this approach it is critical to relate authentication explicitly to access control, because the primary purpose of authentication in a "Need-to-Know" context is precisely to determine (minimal) access rights. Here, we transfer this idea to authentication protocols and introduce "minimise trustworthiness" to balance trust coming from human and computer device domains. We intend to address the maltrust problem from a different aspect.

For DRDs, it is a high-cost and complicated job to deal with unpredictable confusions only depending on computational results. Likewise, given the likelihood of behaving selfishly for the sake of resources, each DRD cannot be simply assumed honest, competent, and willing to perform expensive tasks strictly. So it is unfair to establish trustworthiness from authorities' assurances (due to obvious trust transitivity problem) and it is worse to rely entirely on the results of computations performed by computer devices with no human interaction (another expression of trust transitivity). For instance, when customers withdraw money from an ATM, they cannot ensure (or even verify) that the ATM will implement security policy checking correctly (but interestingly, both banks and customers usually assume ATMs will do so).

---

[6] Regardless of how freely we wish to make resources available, it is dangerous (from the integrity and audit dimensions of security) for users to hold capabilities which they do not even intend to use.

As we pointed out above, human involvement is necessary to the security of pervasive computing. Introducing human knowledge into security protocols has the potential to guide pervasive computer devices to deal with complex security requirements effectively. We have always been inspired by a comment of Mark Weiser, the father of ubiquitous computing, in his well-known paper [31]:

"There is more information available at our fingertips during a walk in the woods than in any computer system, yet people find a walk among trees relaxing and computers frustrating. Machines that fit the human environment, instead of forcing humans to enter theirs, will make using a computer as refreshing as taking a walk in the woods."

6   Fair Authentication in Pervasive Computing

6.1   Leave Strong Entity Authentication Behind

Our belief is that involvement of heavyweight identity-based authentication into protocols is not as secure as initially expected. Instead, such an approach opens attacks unnecessarily in addition to incurring expensive costs. Hence, we ask a logical question, "why not just break the association between identities and authorised resources or access rights?" In other words, "Strong authentication is not always necessary in all circumstances". A similar argument is addressed in Mitchell *et al.*'s paper [20], illustrated with case studies of some protocols in mobile telecommunication systems. Breaking this link can enable us to achieve:

– Privacy and Identity Protection: one desirable consequence is that a raw identity will not be valuable any more so that ID-theft or spoofing make no sense at all. Hence, it protects identity indirectly. Moreover, exclusive identity information in the protocols will satisfy human privacy requirements.
– Data Uncorrelation: another exciting gain is to erase correlation among all input/output data streams with respect to entities. Such uncorrelation makes many active attacks more difficult.

Some may disagree, and argue that unacceptable risks arise by dropping strong authentication from protocols, particularly in the sense of talking to strangers in pervasive computing. It is indeed risky to talk to strangers; however, such risks do not arise from whether protocols are equipped with strong authentication or not. Instead, these risks are coming from the requirements of the applications themselves, e.g. talking to strangers. A similar philosophy applies in human daily life. If Alice trusts Bob whom she has not met or trusted before, then Alice has to risk the possible consequences. Protocols without strong authentication will not necessarily weaken security performance compared with the ones which have. Conversely, they can eliminate the threats accompanied with unnecessary strong authentication.

6.2   "Thinkable Authentication" with Two Level Protocol

Despite the feasibility of leaving traditional entity authentication behind in pervasive computing, distinguishing legitimate users from unauthorised users is still an issue. Here, we propose our second argument "Thinkable Authentication" or human self-determining authentication. Thinking is a distinctive ability in human behaviours, which is unlikely to be exhibited by any computational device[7]. Our "Thinkable" authentication protocols contrast with the traditional "Computable" authentication protocols which involve no distinctively human agency.

Cost and effectiveness are still prior concerns in the design of security protocols [1]. In order to achieve this goal, it is expected to impose necessary tolerances to executed security protocols: for example the majority of communications in the real world can in practice be trusted even though this is not entirely "transparent" trust in most cases of pervasive computing[8]. Thus, two levels of security mechanism are introduced to involve human self-determination knowledge into authentication protocols and allow tradeoff between trustworthiness in both domains (human-human and human-device).

1. A "Plausible (but unreliable) trust" (PT) protocol is used to deal with most legally authorised communications, with protection of current cryptographic encryption techniques (especially using public key cryptography). Since the main purpose of this protocol is to stop passive attacks, trust is plausible, but not reliable for active attacks. This protocol follows basic challenge-response communication between two computational devices (one of which is normally a DRD).

2. A "Reliable trust (acquired through human self-determination)" (RT) protocol is called to achieve higher levels of assurance, with the mandatory interaction of human thought, e.g. monitoring, hearing. This is expected to gain an equivalent outcome to that which strong authentication schemes achieve in traditional computing. Eventually, the protocol's run is completed as success of a "human trust-based decision process" [17].

These two protocols work together to support our "Thinkable Authentication" argument. Some existing protocols can be substantially adapted to the "Thinkable Authentication" hypothesis, e.g. physical contact authentication in Stajano's "Resurrecting Duckling" [26,27,4], entity recognition module [24], authentication starting from weak secrets agreement protocols and applications (which we introduced in section 2), and other reasonable attribute authentications stated in [11]. Potentially, strong computer-based authentication is unnecessary throughout each communication. In fact, frequent recourse to a strong authentication process will be harmful and give cryptanalysis more chances and enough time to collect digital data traffic information.

---

[7] The possibility of devices which can pass the "Turing test" [28] is beyond the scope of this paper, but arguably such devices should be regarded as human users rather than as DRDs from the cyber rights perspective.

[8] "Transparent" trust in this paper means that two entities have established a trust relationship before (e.g. share a secret key) or have been introduced by knowledgeable authorities (e.g. holding corresponding certificates).

6.3   Basic Approach

We will give a simple example of our approach built upon a basic DH-S3P protocol [9], which requires two message exchanges from the PT protocol and human observation from the RT protocol. Consider two company employees A and B who meet each other for the first time in a public conference room, which contains many other people with DRDs. A wishes to send a private company plan m to B via mobile ad-hoc communication. We assume,

DRD_A, DRD_B: two private DRDs held by A and B respectively, which have monitor screens (e.g. PDA) and sufficient computational resources;

k: a weak secret shared by A and B, such as a password which might be reasonably short;

Also, we assume generator g, large prime modulus $q = 2p + 1$ for prime p, and one-way hash function h are publicly known. Random numbers a and b generated by DRDs must be strong (i.e. long enough to be invulnerable to exhaustive search) as well as hard to predict. A full discussion of assumptions as preconditions not specific to the mobile ad-hoc context is given in [9].

- Setup Phase in PT protocol:
    1. A initiates request by inputting k into DRD_A and demands DRD_A (after generating random number a) set up mobile ad-hoc communication with DRD_B;

       $DRD_A \rightarrow DRD_B$: X_A (1)

       where, $X_A = g^a + k \bmod q$;
    2. B inputs k to DRD_B, DRD_B generates random number b and responds;

       $DRD_B \rightarrow DRD_A$: Y_B (2)

       where, $Y_B = g^b + k \bmod q$;
- Marking Phase in RT protocol:

    After the setup phase, neither A nor B has sufficient knowledge to determine whether messages are coming from a genuine human and device or from malicious ones. However, DRD_A and DRD_B both generate a session key s and nonce n from $g^{2ab} \bmod q = (s|n)$. Then, both calculate $n_1 = h\{n\}$ and show n_1 on the screens in some graphic form[9]. Here, we use a slightly different message exchange sequence from the one described in [9]. Mutual authentication is completed only by A and B observing matching images on their screens. Only then, A agrees to send m encrypted under s.

Developing human self-determination into a security protocol allows weak confidentiality to be boosted into strong necessary confidentiality, which is provided by subsequent session key use. In contrast with the original DH-S3P protocol, the security of this approach only partially depends on the continuing secrecy of the shared password k. Devices which behave as "man-in-the-middle" still cannot know the private message m. Admittedly, it is very difficult to guarantee that every human nearby A and B will not peek at the process in order to

---

[9] Perhaps similar to those used in CAPTCHA [29,30]. Such graphics are sufficiently easy for humans to distinguish, but hard for computer devices to spoof, to replace hash function bit-values.

get password k (such cases have been seen in current credit card transactions). Nevertheless, the password in this approach is only a one time shared value and it has a randomness property because it can be any arbitrary value with reasonable length for one-time communication session purpose. Even for one session, it is impractical for an attacker who succeeds in peeking the password to spoof protocols run in the real-time circumstance, because the attacker has not only to peek (human behaviour) the password and get two exchange messages from the wireless channel (computer behaviour) simultaneously for one communication session, but also to peek and reproduce the graphic verifier.

One possibility to attack this approach is to repeat the message (1) and send it to DRD_B, particularly when the communication involves multiple participants (≥ 2) or A and B would like to exchange another company plan $m2$ (after a period) from B to A. In order to block such a replay attack, necessary freshness can be provided using a shared nonce in place of k in the PT protocol.

7   Conclusion

A prototype of fair authentication in pervasive computing has been described, but deeper exploration work is needed for the future. One open issue is the selfishness problem. Most proposed protocols are based upon an assumption that both engaging devices are willing to use up their computational resources to follow protocols, e.g. manual input, doing public key encryption/decryption or hash function calculations. In some sense, communicating devices seem to be selfish in many applications, especially for devices located in a public domain (e.g. public printer in a airport in the example listed in [4,10]). They are unlikely to wish to contribute their resources as much as devices in private domain (e.g. private PDA in the same example as above) which initiate requests. It is not a concern whether these devices in public domain might or might not have sufficient resources. The point is that they might not want to be exhausted and sacrifice their resources for each private request. Thus, authentication protocols are expected to gain the same security performance as most proposed protocols but allow devices in public domain to do less computational tasks. In the meantime, detailed protocols with appropriate countermeasures following the issues introduced in this paper will be specified and implemented in our next step. We anticipate that many conventional authentication protocols will turn out to have "natural" fair counterparts.

In this paper, we introduced a "Thinkable" authentication scheme, which highlights the demand of exploiting human self-determination in pervasive computing instead of traditional storable and delegable authentication scenarios. It follows upon the paradigm shift of pervasive computing. "Fairness" is the faith that motivated us throughout this paper. Essentially, fairness is for the users not for the infrastructures. Only when we realise that the target of security is equivalent to that of fairness, will security protocols be forced to be incorporated from the start of establishing system infrastructure, instead of wrongly being put in afterwards.

References

1. R. Anderson. Why Cryptosystems Fail. *Communications of the ACM*, 37(11): 32 – 40, November 1994.
2. J. Arkko and P. Nikander. Weak Authentication: How to Authenticate Unknown Principals without Trusted Parties. In B. Christianson et al. (Eds.), editor, *Security Protocols: 10th International Workshop, Cambridge, UK*, LNCS2845, pages 5 – 19. Springer - Verlag Berlin Heidelberg, 2004.
3. N. Asokan and P. Ginzboorg. Key Agreement in Ad-hoc Networks. *Computer Communication Review*, (23):1627 – 1637, 2000.
4. D. Balfanz, D. Smetters, P. Stewart, and H. Wong. Talking to Strangers: Authentication in ad-hoc Wireless Networks. In *Symposium on Nework and Distributed Systems Security (NDSS'02)*, February 2002.
5. M. Blaze, J. Feigenbaum, and J. Lacy. Decentralized Trust Management. In *Proc. IEEE Conference on Security and Privacy*, pages 164–173, Oakland, CA, May 1996.
6. M. Burrows, M. Abadi, and R. Needham. A Logic of Authentication. *ACM Transactions on Computer Systems*, 8(1):18–36, 1990.
7. B. Christianson and W. S. Harbison. Why Isn't Trust Transitive? In *Proceedings of the International Workshop on Security Protocols*, LNCS 1189, pages 171–176. Springer-Verlag, 1997.
8. B. Christianson and J. A. Malcolm. Binding Bit Patterns to Real World Entities. In *Proceedings of the 5th International Workshop on Security Protocols*, LNCS 1361, pages 105–113. Springer-Verlag, 1998.
9. B. Christianson, M. Roe, and D. Wheeler. Secure Sessions From Weak Secrets. In *Proceedings of the International Workshop on Security Protocols*, LNCS 3364. Springer-Verlag, 2003.
10. S. Creese, M. Goldsmith, B. Roscoe, and I. Zakiuddin. The Attacker in Ubiquitous Computing Environments: Formalising the Threat Model. In *In Proc. of the 1st International Workshop on Formal Aspects in Security and Trust*, pages 83 – 97, 2003.
11. S. Creese, M. Goldsmith, B. Roscoe, and I. Zakiuddin. Authentication for Pervasive Computing. In D. Hutter et al., editor, *Security in Pervasive Computing 2003*, LNCS 2802, pages 116 – 129. Springer-Verlag Berlin Heidelberg, 2004.
12. C. M. Ellison, B. Frantz, B. Lampson, R. Rivest, B. M. Thomas, and T. Ylonen. SPKI Certificate Theory. Internet rfc 2693, October 1999.
13. C. Gehrmann, C.J. Mitchell, and K. Nyberg. Manual Authentication for Wireless Devices. *Cryptobytes*, 7(1):29 – 37, 2004.
14. L. Gong. *Cryptographic Protocols for Distributed Systems*. Ph.d thesis, University of Cambridge, 1990.
15. D. Hutter, W. Stephan, and M. Ullmann. Security and Privacy in Pervasive Computing State of the Art and Future Directions. In Dieter Hutter et al., editor, *Security in Pervasive Computing 2003, Lecture Notes in Computer Science*, LNCS 2802, pages 285 – 289. Springer-Verlag, 2004.
16. P. A. Karger. *Improving Security and Performance for Capability Systems*. Ph.D thesis, University of Cambridge, 1988.
17. M. Langheinrich. When Trust Does Not Compute – The Role of Trust in Ubiquitous computing. Workshop on Privacy at Ubicomp 2003, October 2003.
18. A. J. Menezes, P. C. van Oorschot, and S. A. Vanstone. *Handbook of Applied Cryptography*. CRC Press, Florida, 1997.

19. S. P. Miller, B. C. Neuman, J. I. Schiller, and J. H. Saltzer. Kerberos Authentication and Authorisation System. Project Athena Technical Plan, section e.2.1, M.I.T, October 1988.
20. C. J. Mitchell and P. Pagliusi. Is Entity Authentication Necessary? In B. Christianson et al., editor, *Security Protocols 2002*, LNCS 2845, pages 20 – 33. Springer - Verlag Berlin Heidelberg, 2004.
21. S. J. Mullender. *Principles of Distributed Operating System Design*. Ph.D thesis, Vrije Universiteit te Amsterdam, 1985.
22. R. M. Needham and M. D. Schroeder. Using Encryption for Authentication in Large Networks of Computers. *Communications of the ACM*, (21(12)):993–999, 1978.
23. M. Roe. *Cryptography and Evidence*. Ph.D thesis, University of Cambridge, 1997.
24. J. M. Seigneur, S. Farrell, C. D. Jensen, E. Gray, and Y. Chen. End-to-End Trust Starts with Recognition. In D. Hutter et al., editor, *Security in Pervasive Computing 2003*, LNCS 2802, pages 130 – 142. Springer - Verlag Berlin Heidelbery, 2004.
25. F. Stajano. Security for Whom? The Shifting Security Assumptions of Pervasive Computing. In *Proceedings of International Security Symposium*, volume 2609. Springer-Verlag, 2002.
26. F. Stajano and R. Anderson. The Resurrecting Duckling: Security Issues for Ad-hoc Wireless Networks. In B. Christianson, B. Crispo, and M. Roe, editors, *Security Protocols, 7th International Workshop Proceedings, Lecture Notes in Computer Science*, LNCS 1296, pages 172 – 194, 1999.
27. F. Stajano and R. Anderson. The Resurrecting Duckling: security issues for ubiquitous computing. *IEEE Computer*, 35(4), April 2002.
28. A.M. Turing. Computing Machinery and Intelligence. *MIND*, 49:433 – 460, 1950.
29. L. von Ahn, M. Blum, N. J. Hopper, and J. Langford. CAPTCHA: Using Hard AI Problems for Security. In *Advances in Cryptology, Eurocrypt 2003*, volume 2656, pages 294 – 311, May 2003.
30. L. von Ahn, M. Blum, and J. Langford. Telling Humans and Computers Apart Automatically. *Communications of the ACM*, 47(2):56 – 60, 2004.
31. M. Weiser. The Computer for the Twenty-First Century. *Scientific American*, 265(3):94 – 104, September 1991.

Cryptanalysis of the Energy Efficient Stream Ciphers SSC2[*]

Yunyi Liu[1], Tuanfa Qin[1,2], Wansun Ni[1], and Shuyi Zhang[1]

[1] State Key Laboratory of Modern Acoustics, Institute of Acoustics,
Nanjing University, Nanjing, Jiangsu 210093, China
[2] College of Computer and Electronic Information, Guangxi University,
Nanning, Guangxi 530004, China

Abstract. The SSC2 is a fast software stream cipher designed for wireless handsets with limited computational capabilities. It is the only one stream cipher which is special designed aim to energy efficient cryptography for wireless sensor networks in recent years open literatures. In this paper, the improved Guess-and-Determine attacks on both LFSR and lagged-Fibonacci half-ciphers of the SSC2 stream cipher are proposed. And some open problems about designing energy efficient stream cipher are discussed.

Keywords: Cryptanalysis, improved Guess-and-Determine attack, wireless sensor networks, fast stream cipher, SSC2.

1   Introduction

The history of stream cipher is a long and wonderful one. There are some notable application-specific standards for stream ciphers, including: A5/1, A5/2 and A5/3 for GSM [1]; RC4 [2] for WEP in IEEE 802.11; f8 [3] function for UMTS/3GPP and E0 [4] for Bluetooth. However, the security of modern stream ciphers is not as well understood as for block ciphers. Most stream ciphers that have been widely spread, like RC4, A5/1, have security weaknesses [5,6]. Also, it is quite remarkable that none of the six submitted stream ciphers submitted to the NESSIE program (BMGL, SNOW, SOBER-t16/32, LEVIATHAN, LILI-128 and RC4 [7]) meets the rather stringent security requirements put forward by NESSIE [8].

Now, the ECRYPT [9] plans to manage and co-ordinate a multi-year effort to identify new stream ciphers suitable for widespread adoption. Also, ISO/IEC 18033-4:2005 [10] just now specifies the following two ways to generate keystream. (I) Mechanisms based on a block cipher: OFB, CTR, and CFB modes of block ciphers; (II) Dedicated keystream generators: MUGI [11] and SNOW 2.0 [12].

It is a real challenge to design the secure modern stream ciphers. It is much more difficulty to design a good stream cipher for the wireless sensor networks

[*] Supported by the support program for 100 Young and Middle-aged Disciplinary Leaders in Guangxi Higher Education Institutions.

M. Burmester and A. Yasinsac (Eds.): MADNES 2005, LNCS 4074, pp. 144–157, 2006.
© Springer-Verlag Berlin Heidelberg 2006

(WSN) [13] or other limited computational capabilities environments. The typical encryption speed of SNOW2.0 is over 3Gbits/sec on a Intel Pentium 4 running at 1.8GHz [12], which is fast enough for the WSN. However, the program of SNOW2.0 is still too large for the WSN, because of the two big tables are applied in the S-box and the multiplication over $GF(2^{32})$ of LFSR. For example, the prototype sensor node deployed in SmartDust project [13], only has 4 MHz 8-bit CPU with 8kb instruction flash, 512 bytes RAM and 512 bytes EEPROM, however there are $256 \times 4 \times 4 = 4096$ bytes tables for S-box and $256 \times 4 = 1024$ bytes tables for each α and α^{-1} of LFSR in SNOW2.0. So, it is necessary to design a fast stream cipher for WSN without complex operations such as multiplication, division, exponentiation and large table look up. To the best of our knowledge, the stream cipher SSC2 is the only stream cipher which is special designed for this target application in the open literatures of recent years.

SSC2 is a stream cipher proposed by Carroll, Chan and Zhang [14]. The cipher is designed for software implementation and is very simple and fast. SSC2 is based on a linear feedback shift register (LFSR) and a lagged-Fibonacci generator (LFG). The LFSR half-cipher consists of a word-oriented LFSR and a nonlinear filter (NLF) which relies on the entire LFSR states by using bit rotations rather than field multiplications. The LFG half-cipher consists of a 17-word LFG and a multiplexor that chooses values from the register of the LFG.

The previous attacks on SSC2 are listed as follows: In the rump session of Crypto 2000, Hawkes and Rose [15] report a distinguish attack on SSC2 based on the correlations between the LSBs of the output words from SSC2, and attack the LFSR half-cipher in isolation by applying the small period of the lagged-Fibonacci generator. Another analysis proposed by Hawkes and Rose [16] found a Guess-and-Determine (GD) attack on the LFSR half-cipher requiring 382 words and around 2^{42} time. Bleichenbacher and Meier [17] found an attack on the entire cipher using about 2^{52} keystream words with about 2^{75} time. Lastly, Hawkes, Quick and Rose [18] found a new faster attack on SSC2. In the faster attack, based on the correlation noted in [15], a fast correlation attack is applied to recovery the initial states of the LFSR, requiring 2^{25} words of 2^{52} keystream words and a few hours of processing time on a 250 MHz Sun UltraSPARC. Then the initial state of the LFG are reconstructed, requiring about 15300 known outputs of the LFG half-cipher and a second of processing on the same computer.

In this paper, a simple but useful strategy is applied to improved the GD attack on LFSR half-cipher in [16] and the GD attack on the lagged-Fibonacci half-cipher in [18]. In this strategy, a portion of the guess values are firstly tested to determine whether the values are wrong or not, to remove the impossible guesses. Then, the surviving guess values can be tested in a faster way. The paper is organized as follows. In Section 2, the stream cipher SSC2 is introduced briefly. The previous and improved GD attacks on LFSR half-cipher of SSC2 are described in Section 3. Then GD attacks on lagged-Fibonacci half-cipher follow in Section 4. Also we address discussion on some open problems about designing faster stream cipher in Section 5 and the concluding remarks are given in Section 6.

2   A Brief Description of SSC2

The stream cipher SSC2 consists of a LFSR filter generator and a lagged-Fibonacci generator, as drawn in Figure 1. In the LFSR filter generator, the word-oriented LFSR has 4 states with each state containing a word. The LFSR generates a new word and shifts out an old word at every clock. Then, the nonlinear filter compresses the 4 states of the LFSR to a word. The lagged-Fibonacci generator is also word-oriented and has 17 states. The word shifted out by the lagged-Fibonacci generator is left cyclic rotated 16 bits and then added to another word selected form the 17 states. The sum is XOR-ed with the word produced by the LFSR filter generator to obtain a keystream word.

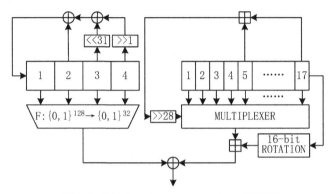

Fig. 1. The keystream generator of SSC2

2.1   The LFSR Filter Generator

For a bit oriented linear feedback shift register (LFSR), a byte-oriented or word-oriented processor needs to spend many clock cycles to perform the bit-shifting and bit-extraction operations. A word-oriented LFSR is designed to make the software implementation more efficient. The bit-oriented characteristic polynomial of the LFSR in SSC2 steam cipher is as

$$p(x) = x(x^{127} + x^{63} + 1). \tag{1}$$

And the word-oriented linear recursion of the LFSR in SSC2 is

$$x_{n+4} = x_{n+2} \oplus (x_{n+1} \ll 31) \oplus (x_n \gg 1), \tag{2}$$

where $x \ll 31$ denotes the zero-fill left-shift 31 bits operation of x, and $\gg$ denotes the zero-fill right-shift operation.

The nonlinear filter in the LFSR filter generator is a memoryless function. Let $(x_{n+3}, x_{n+2}, x_{n+1}, x_n)$ denote the state of the word-oriented LFSR at time n. The output of the LFSR filter generator at time n, is denoted as z'_n. Then, the nonlinear filter is expressed as

$$z'_n = \langle x_{n+3} + (x_n \vee 1) \rangle_{16} + x_{n+2} + c_1(x_n \vee 1) + x_{n+1} \oplus x_{n+2} + c_2 \mod 2^{32}, \tag{3}$$

where $\vee$ denotes bitwise "OR" operation, $\langle A \rangle_{16}$ is a cyclicly shifting A left 16 bits operation. c_1 and c_2 denote the first (line 9) and second (line 15) carry bits in the following ANSI C code of LFSR filter generator.

The LFSR filter generator output can be described in the following ANSI C code:

```
unsigned long R1, R2, R3, R4, outputZ1, temp1, temp2;
int c;
temp1 = R2 ^ (R3<<31) ^ (R4>>1);
R4 = R3;
R3 = R2;
R2 = R1;
R1 = temp1;
temp1 = (R4 | 0x1) + R1;
c = (temp1 < R1);
temp2 = (temp1<<16) ^ (temp1>>16);
if (c) {
    temp1 = (R2 ^ (R4 | 0x1)) + temp2;
}else{
    temp1 = R2 + temp2;}
c = (temp1 < temp2);
outputZ1 = c+ (R3 ^ R2) +temp1;
```

2.2   The Lagged-Fibonacci Generator

The Lagged-Fibonacci has been widely used as random number generators in Monte Carlo simulation, a Lagged-Fibonacci is applied in SSC2 stream cipher, and can be characterized by the following recursion:

$$y_n = y_{n-s} + y_{n-r} \mod M, \ n \ge r, \tag{4}$$

where $s = 5$, $r = 17$ in the SSC2 stream cipher.

The output word of the lagged-Fibonacci generator at time n, denoted z_n'', is given by

$$z_n'' = \langle y_n \rangle_{16} + B[1 + ((y_{n+17} \gg 28) + s_{n+1} \mod 16)] \mod 2^{32}, \tag{5}$$

where B is the state buffer of the Lagged-Fibonacci.

The ANSI C code for z_n'' is listed as follows:

```
unsigned long B[18], outputZ2, temp1, temp2;
int s=5, r=17;
temp1 = B[r];
temp2 = B[s] + temp1;
B[r] = temp2;
if(--r == 0) r = 17;
if(--s == 0) s = 17;
outputZ2=((temp1>>16)^(temp1<<16))+B[(((temp2>>28)+s)&0xf)+1];
```

3   The Guess-and-Determine Attack on the LFSR Filter Generator

3.1   The Previous GD Attack on the LFSR Half-Cipher

First, the GD attack on the LFSR filter generator developed by P. Hawkes and G. Rose [16] is introduced as follows.

Let x_n^* denotes the word x_n with the lest significant bit (LSB) set to one. Then based on (3), the output word z_n' of the LFSR filter generator can be rewritten as

$$
\begin{aligned}
A_n &= x_n^* \boxplus x_{n+3} \to c_{1,n}, & B_n &= \mathrm{SWAP}(A_n), \\
C_n &= B_n \boxplus (x_{n+2} \oplus (c_{1,n} \cdot x_n^*)) \to c_{2,n}, & z_n' &= c_{2,n} \boxplus (x_{n+1} \oplus x_{n+2}) \boxplus C_n,
\end{aligned}
\tag{6}
$$

where $\boxplus$ is addition modulo 2^{32}; $\oplus$ is 32-bit XOR; SWAP denotes the swapping the higher and lower halves of the 32-bit word; $\to$ denotes outputting the carry of the addition.

Then, by dividing the 32-bit words into 16-bit half-words: for example, $x_{n+i} = x_{n+i}^H \parallel x_{n+i}^L$, and by using the modulo 2^{16} (also denoted by $\boxplus$), 16-bit XOR and carries $d_{i,n}$ from the addition of the lower half-words, Eq. (6) can be written as

$$
\begin{aligned}
A_n^L &= x_n^{L*} \boxplus x_{n+3}^L \to d_{1,n}, \\
A_n^H &= x_n^H \boxplus x_{n+3}^H \boxplus d_{1,n} \to c_{1,n}, \\
C_n^L &= A_n^H \boxplus (x_{n+2}^L \oplus (c_{1,n} \cdot x_n^{L*})) \to d_{2,n}, \\
C_n^H &= A_n^L \boxplus (x_{n+2}^H \oplus (c_{1,n} \cdot x_n^H)) \boxplus d_{2,n} \to c_{2,n}, \\
z_n'^L &= c_{2,n} \boxplus (x_{n+1}^L \oplus x_{n+2}^L) \boxplus C_n^L \to d_{3,n}, \\
z_n'^H &= (x_{n+1}^H \oplus x_{n+2}^H) \boxplus C_n^H \boxplus d_{3,n}
\end{aligned}
\tag{7}
$$

$$
z_n'^L = x_n^H \boxplus x_{n+3}^H \boxplus (x_{n+2}^L \oplus (c_{1,n} \cdot x_n^{L*})) \boxplus (x_{n+1}^L \oplus x_{n+2}^L) \boxplus (c_{2,n} \boxplus d_{1,n})
\tag{8}
$$

$$
z_n'^H = x_n^{L*} \boxplus x_{n+3}^L \boxplus (x_{n+2}^H \oplus (c_{1,n} \cdot x_n^H)) \boxplus (x_{n+1}^H \oplus x_{n+2}^H) \boxplus (d_{2,n} \boxplus d_{3,n})
\tag{9}
$$

Based on the above equations, P. Hawkes and G. Rose can recover the states of LFSR by guessing the values of $c_{1,n}$, $(c_{2,n} \boxplus d_{1,n})$ and $(d_{2,n} \boxplus d_{3,n})$ at 10 specifically time n. For each time n, there are 2 possible values for $c_{1,n}$, and 3 possible values each for $(c_{2,n} \boxplus d_{1,n})$ and $(d_{2,n} \boxplus d_{3,n})$. Therefore, the total number of guesses is $(2 \cdot 3^2)^{10} = 2^{41.7}$, so the process complexity of the attack is $c(2^{41.7})$, where $c(N)$ indicates that the complexity is expected to be a small multiple of N. The data complexity of the attack is small, requiring observation of 382 consecutive keystream words for the Guess-and-Determine attack at a single time n.

3.2   The Improved GD Attack on the LFSR Half-Cipher

We notice that the linear relationship is applied in turn from the first LSBs to the seconde LSBs, then to the higher LSBs in the attack of [16]. That is to say, we do not need to guess all values of the carries at the same time. When applying the linear relationship for the first LSBs in (8) and (9),only the first LSBs of

$c_{1,n}$, $(c_{2,n} \boxplus d_{1,n})$ and $(d_{2,n} \boxplus d_{3,n})$ need to be guessed firstly, that is, the total number of guesses is $(2 \times 2 \times 2)^{10} = 2^{30}$. If we can determine most guess values are wrong, then the complexity of future guess values for the second LSBs can be reduced. In this paper, it is found that we can determine about 99.7% of the 2^{30} guesses are incorrect. The proof is described in detail as follows.

Let x_n^0 and x_n^{16} denote the 0th and 16th bit of x_n respectively. Then the linear relationship for the LSBs based on (8) and (9) can be written as

$$
\begin{aligned}
z_n'^0 &= x_n^{16} \oplus x_{n+3}^{16} \oplus (x_{n+2}^0 \oplus (c_{1,n}^0 \cdot x_n^{0*})) \oplus (x_{n+1}^0 \oplus x_{n+2}^0) \oplus (c_{2,n}^0 \oplus d_{1,n}^0) \\
&= x_n^{16} \oplus x_{n+3}^{16} \oplus x_{n+1}^0 \oplus c_{1,n}^0 \oplus (c_{2,n}^0 \oplus d_{1,n}^0),
\end{aligned} \tag{10}
$$

$$
\begin{aligned}
z_n'^{16} &= x_n^{0*} \oplus x_{n+3}^0 \oplus (x_{n+2}^{16} \oplus (c_{1n}^0 \cdot x_n^{16})) \oplus (x_{n+1}^{16} \oplus x_{n+2}^{16}) \oplus (d_{2n}^0 \oplus d_{3n}^0) \\
&= 1 \oplus x_{n+3}^0 \oplus x_{n+1}^{16} \oplus (c_{1,n}^0 \cdot x_n^{16}) \oplus (d_{2,n}^0 \oplus d_{3,n}^0).
\end{aligned} \tag{11}
$$

Based on the characteristic polynomial (1) and the linear recursion (2) of the LFSR, we have

$$
x_{n+127} = x_{n+63} \oplus x_n. \tag{12}
$$

Consider the sets

$$
\begin{aligned}
X &= \{0, 1, 2, 3, 63, 64, 65, 66, 126, 189\}, \\
Y &= \{0, 63, 126, 127, 189, 190, 253, 254, 317, 381\}, \\
Z &= \{0, 1, 2, 3, 63, 64, 65, 66, 126, 127, 128, 129, 130, 189, 190, 191, 192, 193, \\
&\quad\ \ 253, 254, 255, 256, 257, 317, 318, 319, 320, 381, 382, 383, 384\}.
\end{aligned}
$$

Then, by applying the recursion (12), the values of x_{n+j}, $j \in Z$ can be derived from the values of x_{n+i}, $i \in X$. The full deduction is listed as

$$
\begin{array}{lll}
x_{n+127} = x_{n+63} \oplus x_n, & x_{n+128} = x_{n+64} \oplus x_{n+1}, & x_{n+129} = x_{n+65} \oplus x_{n+2}, \\
x_{n+130} = x_{n+66} \oplus x_{n+3}, & x_{n+190} = x_{n+126} \oplus x_{n+63}, & x_{n+191} = x_{n+127} \oplus x_{n+64}, \\
x_{n+192} = x_{n+128} \oplus x_{n+65}, & x_{n+193} = x_{n+129} \oplus x_{n+66}, & x_{n+253} = x_{n+189} \oplus x_{n+126}, \\
x_{n+254} = x_{n+190} \oplus x_{n+127}, & x_{n+255} = x_{n+191} \oplus x_{n+128}, & x_{n+256} = x_{n+192} \oplus x_{n+129}, \\
x_{n+257} = x_{n+193} \oplus x_{n+130}, & x_{n+317} = x_{n+253} \oplus x_{n+190}, & x_{n+318} = x_{n+254} \oplus x_{n+191}, \\
x_{n+319} = x_{n+255} \oplus x_{n+192}, & x_{n+320} = x_{n+256} \oplus x_{n+193}, & x_{n+381} = x_{n+317} \oplus x_{n+254}, \\
x_{n+382} = x_{n+318} \oplus x_{n+255}, & x_{n+383} = x_{n+319} \oplus x_{n+256}, & x_{n+384} = x_{n+320} \oplus x_{n+257}.
\end{array}
$$

Thus, we have

$$
x_{n+j}^m = \bigoplus_{i \in X} \beta_{i,j} x_{n+i}^m, \quad j \in Z, \ i \in X, \ \beta_{i,j} \in \{0,1\}, \ 0 \le m \le 31. \tag{13}
$$

Also note that z_{n+k}, $k \in Y$ rely on the set words $x_{n+j}, j \in Z$, then we obtain the linear relationship for the first LSBs at time $n = 0$ as

$$
z_0'^0 = x_0^{16} \oplus x_3^{16} \oplus x_1^0 \oplus c_0, \qquad z_0'^{16} = 1 \oplus x_3^0 \oplus x_1^{16} \oplus (c_{1,0}^0 \cdot x_0^{16}) \oplus d_0, \tag{14}
$$

$$
z_{63}'^0 = x_{63}^{16} \oplus x_{66}^{16} \oplus x_{64}^0 \oplus c_{63}, \qquad z_{63}'^{16} = 1 \oplus x_{66}^0 \oplus x_{64}^{16} \oplus (c_{1,63}^0 \cdot x_{63}^{16}) \oplus d_{63}, \tag{15}
$$

$$
\begin{aligned}
z_{126}'^0 &= x_{126}^{16} \oplus x_{65}^{16} \oplus x_2^0 \oplus x_{63}^0 \oplus x_0^0 \oplus c_{126}, \\
z_{126}'^{16} &= 1 \oplus x_{65}^0 \oplus x_2^0 \oplus x_{63}^{16} \oplus x_0^{16} \oplus (c_{1,126}^0 \cdot x_{126}^{16}) \oplus d_{126},
\end{aligned} \tag{16}
$$

$$z'^0_{127} = x^{16}_{63} \oplus x^{16}_0 \oplus x^{16}_{66} \oplus x^{16}_3 \oplus x^0_{64} \oplus x^0_1 \oplus c_{127},$$
$$z'^{16}_{127} = 1 \oplus x^0_{66} \oplus x^0_3 \oplus x^0_{64} \oplus x^0_1 \oplus (c^0_{1,127} \cdot (x^{16}_{63} \oplus x^{16}_0)) \oplus d_{127}, \tag{17}$$

$$z'^0_{189} = x^{16}_{189} \oplus x^{16}_{65} \oplus x^{16}_{64} \oplus x^{16}_1 \oplus x^0_{126} \oplus x^0_{63} \oplus c_{189},$$
$$z'^{16}_{189} = 1 \oplus x^0_{65} \oplus x^0_{64} \oplus x^0_1 \oplus x^{16}_{126} \oplus x^{16}_{63} \oplus (c^0_{1,189} \cdot x^{16}_{189}) \oplus d_{189}, \tag{18}$$

$$z'^0_{190} = x^{16}_{126} \oplus x^{16}_{66} \oplus x^{16}_{65} \oplus x^{16}_{64} \oplus x^{16}_2 \oplus x^0_{64} \oplus x^0_{63} \oplus x^0_0 \oplus c_{190},$$
$$z'^{16}_{190} = 1 \oplus x^0_{66} \oplus x^0_{65} \oplus x^0_2 \oplus x^{16}_{64} \oplus x^{16}_{63} \oplus x^{16}_0 \oplus c^0_{1,190}(x^{16}_{126} \oplus x^{16}_{63}) \oplus d_{190}, \tag{19}$$

$$z'^0_{253} = x^{16}_{189} \oplus x^{16}_{126} \oplus x^{16}_{64} \oplus x^{16}_1 \oplus x^{16}_2 \oplus x^0_{126} \oplus x^0_0 \oplus c_{253},$$
$$z'^{16}_{253} = 1 \oplus x^0_{64} \oplus x^0_1 \oplus x^0_2 \oplus x^{16}_{126} \oplus x^{16}_0 \oplus (c^0_{1,253} \cdot (x^{16}_{189} \oplus x^{16}_{126})) \oplus d_{253}, \tag{20}$$

$$z'^0_{254} = x^{16}_{126} \oplus x^{16}_0 \oplus x^{16}_{65} \oplus x^{16}_3 \oplus x^{16}_2 \oplus x^0_{63} \oplus x^0_1 \oplus x^0_0 \oplus c_{254},$$
$$z'^{16}_{254} = 1 \oplus x^0_{65} \oplus x^0_3 \oplus x^0_2 \oplus x^{16}_{63} \oplus x^{16}_1 \oplus x^{16}_0 \oplus c^0_{1,254}(x^{16}_{126} \oplus x^{16}_0) \oplus d_{254}, \tag{21}$$

$$z'^0_{317} = x^{16}_{189} \oplus x^{16}_{63} \oplus x^{16}_{66} \oplus x^{16}_{65} \oplus x^{16}_{64} \oplus x^{16}_1 \oplus x^0_{126} \oplus x^0_{64} \oplus x^0_{63} \oplus c_{317},$$
$$z'^{16}_{317} = 1 \oplus x^0_{66} \oplus x^0_{65} \oplus x^0_{64} \oplus x^0_1 \oplus x^{16}_{126} \oplus x^{16}_{64} \oplus x^{16}_{63} \oplus c^0_{1,317}(x^{16}_{189} \oplus x^{16}_{63}) \oplus d_{317}, \tag{22}$$

$$z'^0_{381} = x^{16}_{189} \oplus x^{16}_{126} \oplus x^{16}_{63} \oplus x^{16}_0 \oplus x^{16}_{66} \oplus x^{16}_{64} \oplus x^{16}_3 \oplus x^{16}_2 \oplus x^{16}_1$$
$$\oplus x^0_{126} \oplus x^0_{64} \oplus x^0_1 \oplus x^0_0 \oplus c_{381},$$
$$z'^{16}_{381} = 1 \oplus x^0_{66} \oplus x^0_{64} \oplus x^0_3 \oplus x^0_2 \oplus x^0_1 \oplus x^{16}_{126} \oplus x^{16}_{64} \oplus x^{16}_1 \oplus x^{16}_0$$
$$\oplus (c^0_{1,381} \cdot (x^{16}_{189} \oplus x^{16}_{126} \oplus x^{16}_{63} \oplus x^{16}_0)) \oplus d_{381}, \tag{23}$$

where $c_n = c^0_{1,n} \oplus c^0_{2,n} \oplus d^0_{1,n}$, $d_n = d^0_{2,n} \oplus d^0_{3,n}$.

We find that x^0_{n+189} never appears in equations (14) to (23). So, if all the 30 carries are guessed and all z' are known, 20 linear equations of 19 variables are obtained. The solution is obtained by applying Gaussian elimination, and we can determine if at least one of the guess values is wrong. For example:

The random initial values of LFSR are: $x_0 = 0xfcc6b54a$, $x_1 = 0x4a14caa8$, $x_2 = 0x0d5aba4c$, $x_3 = 0x5f2ac355$. The LFSR filter output are

$$z'^0_0 = 1, \ z'^{16}_0 = 0, \ z'^0_{63} = 0, \ z'^{16}_{63} = 1, \ z'^0_{126} = 0, \ z'^{16}_{126} = 1, \ z'^0_{127} = 0, \ z'^{16}_{127} = 1,$$
$$z'^0_{189} = 0, \ z'^{16}_{189} = 0, \ z'^0_{190} = 1, \ z'^{16}_{190} = 1, \ z'^0_{253} = 1, \ z'^{16}_{253} = 0, \ z'^0_{254} = 0, \ z'^{16}_{254} = 0,$$
$$z'^0_{317} = 0, \ z'^{16}_{317} = 0, \ z'^0_{381} = 0, \ z'^{16}_{381} = 0.$$

If the guess values (random choice) are

$$
\begin{array}{lll}
c^0_{1,0} = 0, & (c^0_{2,0} \oplus d^0_{1,0}) = 0, & (d^0_{2,0} \oplus d^0_{3,0}) = 1, \\
c^0_{1,63} = 1, & (c^0_{2,63} \oplus d^0_{1,63}) = 0, & (d^0_{2,63} \oplus d^0_{3,63}) = 0, \\
c^0_{1,126} = 1, & (c^0_{2,126} \oplus d^0_{1,126}) = 1, & (d^0_{2,126} \oplus d^0_{3,126}) = 1, \\
c^0_{1,127} = 0, & (c^0_{2,127} \oplus d^0_{1,127}) = 0, & (d^0_{2,127} \oplus d^0_{3,127}) = 1, \\
c^0_{1,189} = 1, & (c^0_{2,189} \oplus d^0_{1,189}) = 0, & (d^0_{2,189} \oplus d^0_{3,189}) = 1, \\
c^0_{1,190} = 0, & (c^0_{2,190} \oplus d^0_{1,190}) = 0, & (d^0_{2,190} \oplus d^0_{3,190}) = 0, \\
c^0_{1,253} = 0, & (c^0_{2,253} \oplus d^0_{1,253}) = 1, & (d^0_{2,253} \oplus d^0_{3,253}) = 0, \\
c^0_{1,254} = 0, & (c^0_{2,254} \oplus d^0_{1,254}) = 1, & (d^0_{2,126} \oplus d^0_{3,254}) = 1, \\
c^0_{1,317} = 0, & (c^0_{2,317} \oplus d^0_{1,317}) = 0, & (d^0_{2,126} \oplus d^0_{3,317}) = 1, \\
c^0_{1,318} = 1, & (c^0_{2,318} \oplus d^0_{1,318}) = 1, & (d^0_{2,126} \oplus d^0_{3,318}) = 1.
\end{array}
$$

Let, $F \cdot X = b$, $X = [\ x_0^0, x_0^{16},\ x_1^0, x_1^{16},\ x_2^0, x_2^{16},\ x_3^0, x_3^{16},\ x_{63}^0, x_{63}^{16},\ x_{64}^0, x_{64}^{16},$ $x_{65}^0, x_{65}^{16},\ x_{66}^0, x_{66}^{16},\ x_{126}^0, x_{126}^{16},\ x_{189}^0, x_{189}^{16}]'$. Then after applying the Gaussian elimination to this equations, we have

$$F = \begin{bmatrix}
1,0,0,0,0,1,0,0,1,0,0,0,0,1,0,0,0,0,0,0 \\
0,1,0,0,0,0,0,0,0,0,0,0,0,0,0,0,0,0,0,0 \\
0,0,1,0,0,0,0,0,0,0,1,0,0,1,0,0,0,0,0,0 \\
0,0,0,1,0,0,0,0,1,0,0,0,0,1,1,0,1,0,0,0 \\
0,0,0,0,1,0,0,0,0,0,0,0,1,0,0,0,0,0,0,0 \\
0,0,0,0,0,0,0,0,0,0,0,0,0,0,0,0,0,0,0,0 \\
0,0,0,0,0,0,1,0,1,0,0,0,0,1,1,0,1,0,0,0 \\
0,0,0,0,0,0,0,0,1,0,0,0,0,1,0,0,1,0,0,0 \\
0,0,0,0,0,0,0,0,0,0,0,0,0,0,0,0,0,0,0,0 \\
0,0,0,0,0,0,0,0,0,0,1,0,0,0,0,0,0,0,0,0 \\
0,0,0,0,0,0,0,0,0,0,0,1,0,0,0,0,1,0,0,0 \\
0,0,0,0,0,0,0,0,0,0,0,0,1,0,0,1,0,0,0,0 \\
0,0,0,0,0,0,0,0,0,0,0,0,0,0,0,0,0,0,0,0 \\
0,0,0,0,0,0,0,0,0,0,0,0,0,0,0,0,0,0,0,0 \\
0,0,0,0,0,0,0,0,0,0,0,0,0,0,0,0,0,0,0,0 \\
0,0,0,0,0,0,0,0,0,0,0,0,0,0,0,0,0,0,0,0 \\
0,0,0,0,0,0,0,0,0,0,0,0,0,0,0,0,1,0,0,0 \\
0,0,0,0,0,0,0,0,0,0,0,0,0,0,0,0,0,0,0,0 \\
0,0,0,0,0,0,0,0,0,0,0,0,0,0,0,0,0,0,0,1
\end{bmatrix}, \quad b = \begin{bmatrix} 0 \\ 1 \\ 0 \\ 1 \\ 1 \\ 0 \\ 1 \\ 0 \\ 1 \\ 0 \\ 1 \\ 1 \\ 1 \\ 0 \\ 1 \\ 0 \\ 0 \\ 1 \\ 1 \end{bmatrix}.$$

In the above equations, the coefficients in 13th, 14th, 16th and 19th rows of F are all zeros, while the output values still are 1 in the corresponding rows of b. For this situation we can determine that the guess is wrong. We are surprised that we can determine about 99.7% of 2^{30} times of guess is incorrect in this way. For example, when $x_0 = 0x070ad440$, $x_1 = 0x0ac8fa3e$, $x_2 = 0x036d67c1$, $x_3 = 0x208ce468$ and the first LSBs of all carries are guessed from (14) to (23), that is total $2^{30} = 1073741824$ times of guess, according to the method described on above, we know that $1070190080 \approx 2^{29.995}$ times of guess must be wrong, and $3551744 \approx 2^{21.8}$ times of guesses need to be determined later.

Processing the attack on the second LSBs in turn as [16], we can finish the attack on the LFSR filter generator. Note that we have excluded most of guess values, and only few candidates can survival. So the remaining guesses will be easy, and the process complexity of the Guess-and-Determine attack on the LFSR filter generator in the SSC2 stream cipher can reduce to about $c(2^{30})$.

4   Attack on Lagged-Fibonacci Generator

4.1   Redescribe the Lagged-Fibonacci Generator

To describe the attack on the lagged-Fibonacci generator conveniently, the lagged-Fibonacci generator is reexpressed as Figure 2 based on the definition in (5) and the ANSI C code of the generator.

Let $B_n, n = 1, 2, \cdots$, be the states of the lagged-Fibonacci, Z_n'' be the generator output in time n, we have

$$Z_n'' = \text{SWAP}(B_n) \boxplus B_{k_n}, \quad n = 1, 2, \cdots, k_n \in [n+1, n+17] \qquad (24)$$

where k_n is the index for selecting B_{k_n} from the candidate states $B_{n+1,\cdots,n+17}$ at time n. The index k_n is depend on B_{n+17} and the time n.

The new equivalent ANSI C code is listed as

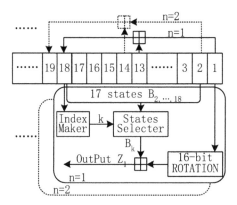

Fig. 2. The lagged-Fibonacci generator

```
unsigned long DataLen=1000;      //The length of output.
unsigned long* B;                //The states buffer.
B = new unsigned long[DataLen+18];
for(int I=1;I<18;I++)            //Set the initial values.
    B[I]=rand()*rand()*rand();   //Random values for test.
unsigned long n;                 //Time variable.
unsigned long Z;                 //Generator output.
unsigned long k;                 //Selection index.
for(n=1;n<DataLen;n++)
{
    B[n+17] = B[n+12] + B[n];    //Lagged-Fibonacci.
    //Index Maker.
    k = (17 + 5 - n%17)%17;
    if(k == 0) k = 17;
    k = (17- (((B[n+17]>>28)+k)%16))%17;
    while(k <= n)
        k = k + 17;
    //Generator output
    Z = ((B[n]>>16) ^ (B[n]<<16)) + B[k];
}
delete[] B;                      //Release the buffer.
```

4.2   The Strategy of GD Attack on the Lagged-Fibonacci Generator

After the lagged-Fibonacci generator is redescribed, the analysis of the structure becomes easier. The attack strategy on the lagged-Fibonacci generator is the same as applied in above subsection. That is, when guessing some values, we do not need to determine whether they are correct or not directly, on the contrary, we make sure whether they are wrong or not. This strategy is simple but powerful. If we know that most of the guess values must be wrong, then we can determine the survival guess values (those maybe wrong values) later in a faster way. The detail description is listed as follows.

If we guess $k_1 = 13$ and $k_{13} = 18$, according to (24), get

$$
\begin{aligned}
Z_1''^L &= B_1^H \boxplus B_{13}^L \;\rightarrow\; d_1, & Z_1''^H &= B_1^L \boxplus B_{13}^H \boxplus d_1, \\
Z_{13}''^L &= B_{13}^H \boxplus B_{13}^L \boxplus B_1^L \;\rightarrow\; d_{13}, & Z_{13}''^H &= B_{13}^L \boxplus B_{13}^H \boxplus B_1^H \boxplus d_{13},
\end{aligned}
\tag{25}
$$

where $d_1 = 0, 1$ and $d_{13} = 0, 1, 2$.

Solve the set of linear equations (25) to obtain

$$
\begin{aligned}
B_{13}^H &= (Z_{13}''^H - Z_1''^L - d_{13} + 2^{16}) \bmod 2^{16}, \\
B_{13}^L &= (Z_{13}''^L - Z_1''^H + d_1 + 2^{16}) \bmod 2^{16}, \\
B_1^H &= (Z_1''^L - B_{13}^L + 2^{16}) \bmod 2^{16}, \\
B_1^L &= (Z_1''^H - B_{13}^H - d_1 + 2^{16}) \bmod 2^{16}.
\end{aligned}
\tag{26}
$$

By guessing all the possible values of d_1 and d_{13}, the solutions of (26) are obtained. That is, for each guess, B_{13}^H, B_{13}^L, B_1^H and B_1^L are recovered, then the guess values of d_1 and d_{13} can be determined to be correct or not.

After B_{13}^H, B_{13}^L, B_1^H and B_1^L are solved, B_{18} is obtained. Then the new states select index k_1' is figured out. If k_1' is unequal to the guess index k_1, we can determine that the guess must be incorrect, that is to say, k_1 equals to 13 and k_{13} equals to 18 at the same time is impossible.

For example: If $B_1 = 0x3c4ea7eb$ and $B_{13} = 0x2178cc40$, then based on the ANSI C code of the lagged-Fibonacci generator, we have $B_{18} = 0x5dc7742b, Z_1'' = 0xb0d1cab2$, $Z_{13}'' = 0xd194f3c3$, $k_1 = 8$, and $k_{13} = 16$.

Turn to the situation of the attack on the lagged-Fibonacci generator. The target of the attack is to determine the guess: $k_1 = 13$ and $k_{13} = 18$ whether must be wrong or may be correct. Suppose we have $Z_1'' = 0xb0d1cab2$ and $Z_{13}'' = 0xd194f3c3$ and guess $k_1 = 13$ and $k_{13} = 18$. Then solutions of (26) are obtained as listed in Table 1.

Table 1. The solutions of Eq. (26)

Guess values		Recover values					
d_1	d_{13}	B_1	B_{13}	d_1'	d_{13}'	B_{18}	k_1'
0	0	$0x87c0a9ef$	$0x06e242f2$	0	0	$0x8ea2ece1$	4
0	1	$0x87c0a9f0$	$0x06e142f2$	0	0	$0x8ea1ece2$	4
0	2	$0x87c0a9f1$	$0x06e042f2$	0	0	$0x8ea0ece3$	4
1	0	$0x87bfa9ee$	$0x06e242f3$	0	0	$0x8ea1ece1$	4
1	1	$0x87bfa9ef$	$0x06e142f3$	0	0	$0x8ea0ece2$	4
1	2	$0x87bfa9f0$	$0x06e042f3$	0	0	$0x8e9fece3$	4

From Table 1, we learn to know that the carries of (26) must be $d_1 = d_{13} = 0$. Then we have $k_1' \neq k_1$, it means that the guess must be wrong, and $k_1 = 13$ with $k_{13} = 18$ at the same time is impossible.

In the same way, we can check all the guesses: $k_n = n+12$ with $k_{n+12} = n+17$. From this we can determine about 93.8% of guess must be wrong. There are 16 possible values for k_n at a time n, so $k_n = n + 12$ with $k_{n+12} = n + 17$ occurs every $16^2 = 256$ words on average. After the check, only few candidates can survive, and the guess occurs every $256 \times 0.062 \approx 16$ words of the survivals on average.

4.3 The Improved GD Attack on Lagged-Fibonacci Generator

Notice that, the lagged-Fibonacci generator output (Z_n'', Z_{n+12}'') with $k_n = n + 12, k_{n+12} = n + 24$ are called *good pairs* in the attack proposed by Hawkes, Quick and Rose [18]. The initial states of lagged-Fibonacci can be derived from the good pairs by guessing the all carry bits and identifying 17 good pairs.

In fact, the output pairs (Z_n'', Z_{n+12}'') with $k_n = n + 12, k_{n+12} = n + 17$, or with $k_n = n + 17, k_{n+12} = n + 17$, or with $k_n = n + 17, k_{n+12} = n + 24$, can also be considered as other kinds of good pairs. In the similar way as in [18], the initial states of lagged-Fibonacci can be derived from those new good pairs. For example, if 17 good pairs with $k_n = n + 12, k_{n+12} = n + 17$ are obtained, then the 17 states of lagged-Fibonacci can be solved as solving the Eq. (26). Also notice that the carries are determined automatic in solving the Eq. (26), so the problem of guessing the 17 carry bits in [18] can be removed. Then the remaining problems in [18] are identifying the good pairs. Before applying the identifying trick in [18], we can check those pairs by the method described on above subsection to eliminate the most bad pairs.

So, we can improve the Guess-and-Determine attack on lagged-Fibonacci generator of [18] in the following three ways: (I) Auto-determine the carry bits of solving equations; (II) Removing the most of bad pairs to make the identifying good pairs faster; (III) More types of good pairs to reduce the number of outputs required for the attack on lagged-Fibonacci generator.

5   Open Problems About Designing Fast Stream Cipher

5.1 Summarize the Weaknesses of SSC2

Based on the previous attacks in [15,16,17,18] and the improved attack in this paper, the weaknesses of the stream cipher SSC2 can be listed as follows.

(I). The small period of the lagged-Fibonacci: In SSC2, $\pi = 17 \cdot 2^{31} \cdot (2^{17} - 1)$ is the period of the lagged-Fibonacci generator half-cipher. If two segments of output stream π apart are exclusive-ored together, the contributions from LFG half-cipher cancel out leaving the exclusive-or of two filtered LFSR streams to be analyzed [18].

(II). The sparse taps of LFSR: the fast correlation attack in [18] is aided greatly by the fact that the feedback polynomial of the LFSR has only 3 taps. Also, the GD attack in [16] and in this paper are depended on the 3 taps word-oriented linear recursion $x_{n+127} = x_{n+63} \oplus x_n$. In fact, Meier and Staffelbach observed in [19] "any correlation to an LFSR with less than 10 taps should be avoided".

(III). The simple SWAP function: Much of the analysis of SSC2 is based on dividing the 32-bit words into two 16-bit blocks: $A = A^H \| A^L$. Also, the simple SWAP function is the source of the linear correlation $P(z_n'^0 = x_{n+1}^{15} \oplus x_{n+1}^{16} \oplus x_{n+2}^{31} \oplus x_{n+3}^{0} \oplus x_{n+4}^{16}) = 5/8$.

(IV). The weak frame key Scheduling: The frame key generation algorithm of SSC2 do not satisfy the property that it is difficult to gain information about one frame key from another frame key [17].

5.2 Designing a Fast Stream Cipher

The first step of designing a fast stream cipher is to find a pseudo-random number generator (PRNG) with good statistical properties. There are many kinds of PRNG in the world, such as, the linear congruential PRNG, Discrete logarithm PRNG [20], quadratic residuosity PRNG [21], one way functions PRNG [22], Chaos PRNG [23], LFSR PRNG [24], and Fibonacci PRNG [25]. With the exception of the LFSR PRNG, those PRNGs are not suitable for fast modern stream cipher because they have some weakness such as small period of high computed complexity. In fact, the PRNG based on LFSR is applied in modern stream cipher widely. According to [19], it is preferable for the number of taps of a LFSR to equal about half of the LFSR bit length. In particular if the number of feedback taps is small, then it is possible to implement the word-oriented software LFSR in a fast way without table look up operation. However, if the number of taps becomes bigger, the table look up operation is applied in the word-oriented software implementation, such as the LFSRs in SOBER and SNOW stream ciphers. As soon as the table look up operation is applied, especially for the large table, the stream cipher becomes slow. So the first problem of designing fast software stream cipher for WSN is to design a word-oriented height taps LFSR with small or no table look up operation.

The second problem of designing a fast stream cipher is that of designing a secure nonlinear filter (NLF). The NLF output should be balanced to avoid the fast correlation attacks. Also the NLF should be heightly nonlinear to avoid the algebraic attacks. The most difficulty restriction is that we can not apply the strong S-box of block cipher, because the table is still too large for WNS. So, the SWAP function and the multiplexer are the two efficient functions leave to us to construct a strong fast nonlinear filter. There is a possible countermeasure for the SWAP function of SSC2. That is to change the normal SWAP function: $SWAP(A) = (A << 16) \oplus (A >> 16)$, to a irregular format: $SWAP(A, k) = (A << k) \oplus (A >> (32 - k))$, where k depends on the states of LFSR. Also, if the range of the select index in multiplexor is bigger (like the one of RC4), it will make the stream cipher stronger.

The last problem is the key Scheduling for stream cipher. Most stream ciphers just load the master key and IV into the initial states of LFSR, then run the cipher for several hundreds clock without output any keystream to finish the rekey operation. This kind of rekey Scheduling is very useful, the security of the key Scheduling just depend on the stream cipher itself. Also frequent rekey as E0 stream cipher will avoid the know plaintext attacks effectively.

6   Conclusion

In this paper, a simple but useful strategy is applied to improved the Guess-and-Determine attacks on SSC2 stream cipher. In this strategy, one firstly determines whether the guess values are wrong or not, to remove the most impossible guesses. Then, the surviving guess values can be tested in a faster way. By applying the new strategy, we can reduce complexity of the Guess-and-Determine attack on the LFSR half-cipher in [16] from $c(2^{42})$ to about $c(2^{30})$. The GD attack on the lagged-Fibonacci half-cipher in [18] is improved by the same trick. At last, some open problems about designing energy efficient stream cipher are discussed.

Acknowledgement

We are very grateful to Philip Hawkes and Greg Rose for their kindly support, helpful suggestions and carefully reading to improve the quality of the paper. Also thanks to the anonymous reviewers for their useful comments.

References

1. M. Walker and T. Wright. Security. In F. Hillebrand, editor, GSM and UMTS: The creation of global mobile communication, pages 385-406. John Wiley & Sons, 2002.
2. B. Schneier. Appled cryptography: Protocols, algorithms and source code in C. John Wiley & Sons Inc., New York, 2nd edition, 1996.
3. ETSI. 3GPP TS 35.201. Specification of the 3GPP Confidentiality and Integrity Algorithms; Document 1: f8 and f9 Specification, June 2002.
4. Bluetooth SIG. Speciffcation of the Bluetooth System 2.0. Available at Htttp://www.bluetooth.com, Accessed January 2005.
5. Thomas Johansson. Analysis and Design of Modern Stream Ciphers (Invited Paper). IMA 2003, Lecture Notes in Computer Science, Vo. 2898, page: 66, 2003.
6. C. J. Mitchell and A. W. Dent, International standards for stream ciphers: A progress report. SASC-The State of the Art of Stream Ciphers, Novotel Brugge Centrum, Brugge, Belgium, 14th-15th October 2004.
7. NESSIE. NESSIE security report. NES/DOC/ENS/WP5/D20/2, February 19, 2003.
8. NESSIE. NESSIE PROJECT ANNOUNCES FINAL SELECTION OF CRYPTO ALGORITHMS. February 27, 2003.
9. http://www.ecrypt.eu.org
10. ISO/IEC FDIS 18033-4:2005. Information technology – Security techniques – Encryption algorithms – Part 4: Stream ciphers. 2005.
11. S. Furuya, D. Watanabe, Y. Seto, and K. Takaragi. Integrity-aware mode of stream cipher. IEICE Trans. Fundamentals, E85-A:58-65, 2002.
12. P. Ekdahl and T. Johansson. A new version of the stream cipher SNOW. SAC 2002, Lecture Notes in Computer Science, vol. 2595, pages: 47-61, 2003.
13. K. S. J Pister, J. M. Kahn, B. E. Boser, Smart Dust: Wireless networks of milimeter-scale sensor nodes. Highlight Article in 1999 Electronics Research Laboratory Research Summary, 1999

14. C. Carroll, A. Chan, and M. Zhang. The software-oriented stream cipher SSC-II. FSE 2000, Lecture Notes in Computer Science, vol. 1978, pages: 39-56, 2000.
15. P. Hawkes and G. Rose. Correlation cryptanalysis of SSC2. Presented at the Rump Session of Crypto 2000.
16. P. Hawkes and G. Rose. Exploiting multiples of the connection polynomial in word-oriented stream ciphers. Advances in Cryptology, AsiaCrypto2000, Lecture Notes in Computer Science, vol. 1976, pages 302-316, 2000.
17. D. Bleichenbacher and W. Meier. Analysis of SSC2. FSE 2001, Lecture Notes in Computer Science, vol. 2355, pages 219-233, 2001.
18. P. Hawkes, F. Quick and G. Rose. A practical cryptanalysis of SSC2. Selected Areas in Cryptography 2001, Lecture Notes in Computer Science, vol. 2259, pages: 27-37, 2001.
19. W. Meier and O. Staffelbach. Fast correlation attacks on certain stream ciphers. Journal of Cryptology, 1(3): 159-176, 1989.
20. M. Blum and S. Micall. How to generate cryptographically strong sequences of pseudo-random bits. Procceedings of 25th IEEE Symposium on Foundations of Computer Science, New York, pages: 850-864, 1982.
21. L. Blum, M. Blum and M. Shub. A simple unpredictable pseudo-random number generator. Siam J. on Computing, pages: 364-393, 1986.
22. A.C. Yao. Theory and applications of trapdoor functions. Proceedings of the 25th IEEE Symposium on Foundations of Computer Science, IEEE, New York, pages: 80-91, 1982.
23. Ljupco Kocarev, Goce Jakimoski and Zarko Tasev. Chaos and Pseudo-Randomness. Chaos Control: Theory and Applications, Lecture Notes in Control and Information Sciences, vol. 292, pages: 247-264, 2004.
24. H. Beker and F. Piper. Cipher Systems. Johan Wiley, 1982.
25. F. James. A review of pesudo-random number generators. Computer Physis Communications, vol. 60, pages: 329-344, 1990.

ARMS: An Authenticated Routing Message in Sensor Networks

Suk-Bok Lee and Yoon-Hwa Choi

Department of Computer Engineering
Hongik University 121-791 Seoul, Korea
{sblee, yhchoi}@cs.hongik.ac.kr

Abstract. In wireless sensor networks, a sensor node broadcasts its data (such as routing information, beacon messages or meta-data) to all its neighbors, which is called *local broadcast*. A general case for a sensor node to use a local broadcast is to advertise its routing information. Considering that sensor networks are vulnerable to a variety of attacks and current routing protocols are insecure, a sensor node's broadcast message should be authenticated by all its neighbors. Unfortunately, the previous work on broadcast authentication in sensor networks mainly concentrates on broadcast messages from a base station which has greater capabilities, not from a sensor node. Those schemes' properties are not appropriate for broadcast authentication of sensor node's routing messages. In this paper, we present ARMS, a protocol for broadcast authentication of sensor node's routing messages. It requires only a small memory space, authenticates routing messages without delay (thus, no buffering is needed), needs no time synchronization among sensor nodes, and mitigates the effect of packet loss. These ARMS' properties are suitable for a sensor node to broadcast an authenticated routing message.

1 Introduction

There have been a lot of researches in the areas of wireless sensor networks. Those are mainly about energy-efficiency with sensor node's severe hardware constraints. In real situation, however, security is also an important aspect in sensor networks where sensor nodes are deployed in harsh environments. A variety of attacks could be launched such as node capture, physical tampering, denial of service, etc [11].

Recently, researchers in sensor networks have actively investigated security problems and proposed security mechanisms that protect against malicious attackers who do eavesdropping, injecting malicious packets, replaying old messages, etc. While key-management schemes for establishing pairwise keys among neighboring nodes [1, 2, 7, 12] and security mechanisms for node-to-node secure communication [6][12] such as link layer encryption and authentication mechanisms in sensor networks are well defined, many other security issues need further investigation. Especially, current routing protocols for sensor networks are insecure and they are highly vulnerable to node capture attacks [5]. By node capture

M. Burmester and A. Yasinsac (Eds.): MADNES 2005, LNCS 4074, pp. 158–173, 2006.

attacks, turning a legitimate node into a malicious node and injecting malicious routing information that leads to routing inconsistencies, attackers can take over some part of network or whole network topology. It turns out to be relatively easy to compromise a node [3], which is to extract cryptographic keys from a captured node and to make malicious code running for the attacker's purpose.

To protect the network from the malicious nodes that inject forged routing information, secure routing protocols should have a mechanism that detects and isolates these compromised nodes from the network. To do that, we first should have the capability to authenticate routing messages, which prevents a malicious node from spoofing its identity during routing advertisements. Since routing advertisements are locally broadcast messages (i.e. one-to-many), without an asymmetric mechanism, any compromised receiver could forge messages from the sender. During routing advertisements a compromised node should not be allowed to forge its identity to impersonate other legitimate nodes (e.g. any of its neighbor nodes). If a compromised node has the ability to impersonate other legitimate nodes during routing advertisement, it would be very difficult or impossible for secure routing protocols to distinguish a compromised node from legitimate nodes impersonated. Thus, the mechanism, which guarantees that routing advertisement messages were really sent from the claimed node, could be considered as a basis for routing security. Armed with this fundamental mechanism, researchers will be able to design a secure routing protocol that detects and isolates a compromised node injecting inconsistent routing messages.

In this paper, we present ARMS (An Authenticated Routing Message in Sensor Networks), a protocol for broadcast authentication of sensor node's routing messages. It requires only a small memory space, authenticates routing messages without delay (thus, no buffering is needed), needs no time synchronization for broadcast messages among sensor nodes, and mitigates the effect of packet loss. These ARMS' properties make it suitable for a sensor node to broadcast an authenticated routing message.

In Section 2, we briefly introduce an overview of related work. Background is given in Section 3. In Section 4, our protocol, ARMS is presented and described in detail. An evaluation of ARMS is given and discussed in Section 5. We draw conclusions in Section 6.

2   Related Work

Recently, security problems in sensor networks have been investigated by several researchers [13, 5, 9, 12, 8]. Wood *et al.* [13] have listed each layer's vulnerabilities in sensor networks. Several attacks are possible to be launched in each layer. Karlof *et al.* [5] have especially focused on security in currently proposed routing protocols in sensor networks. Their analysis reveals that current routing protocols are insecure. Particularly, attacks launched from insiders (which means the adversaries compromise the legitimate nodes and make them affect the routing topology adversely) are most dangerous, and they leave it as an open problem to design secure routing protocols.

Perrig *et al.* [9] have proposed TESLA protocol for broadcast authentication which is intended for other than sensor networks. It is an efficient authentication scheme for multicast streams over lossy channels. They [10] have improved TESLA protocol to allow immediate authentication by replacing receiver-side buffering with sender-side buffering. Although those versions of TESLA have nice properties, they cannot be used in sensor networks for several reasons like high overhead of computing digital signatures.

Perrig *et al.* [12] have also proposed μTESLA for broadcast authentication in sensor networks. Since a resource-constraint sensor node cannot afford high commutation overhead of asymmetric cryptography mechanisms, digital signature schemes is not suitable for broadcast authentication in sensor networks. μTESLA deals successfully with this problem by delaying the disclosure of symmetric keys. To work properly, loosely time synchronization is required between the base station and sensor nodes. The base station generates one-way key chain before it starts broadcasting to sensor nodes. Sensor nodes can authenticate the received broadcast message when the base station discloses the corresponding delayed key.

Liu *et al.* [8] have proposed *multilevel* μTESLA, which is based on μTESLA and extends the capabilities of μTESLA. While *multilevel* μTESLA keeps nice properties of μTESLA, it eliminates the initial unicast message transmissions between the base station and sensor nodes to improve scalability to large networks.

μTESLA and *multilevel* μTESLA are efficient authentication schemes for the base station's broadcast messages. However, since both are designed for authentication of the base station's broadcast message, they are not suitable for authentication of a sensor node's broadcast routing messages for several reasons to be addressed in Section 3.4. Even though a sensor node can broadcast authenticated messages via the base station to other sensor nodes in [12], it's impractical for advertising sensor nodes' routing messages. Thus, a new broadcast authentication mechanism for a sensor node's routing messages is needed.

3   Background

3.1   Network Assumption

We assume that each pair of neighboring nodes establishes a secure communication channel by agreeing on a unique secret key between two neighboring nodes. There are several proposed schemes we can employ for establishing a unique secret key between two neighboring nodes [1, 2, 7, 12]. Any particular scheme chosen has nothing to do with the operation of ARMS. Depending on the schemes we choose, some pairs of neighboring nodes might fail to agree on secret keys. However, sensing data will be forwarded only through secure communication channels, so a pair of neighboring nodes which eventually failed to agree on a secret key has to be regarded as no more a neighborhood.

3.2   Adversarial Model

The adversary is assumed to have ability to capture a legitimate node and turn it into a malicious node, which is to extract cryptographic keys from a captured node and to make malicious code running for the attacker's purpose. Having an intention of taking over some part of network or whole network topology, the adversary could inject malicious routing information that leads to routing inconsistencies. During routing advertisements a compromised node can advertise forged routing information (e.g. an extremely high quality route to a base station) to attract a large amount of network traffic [5]. This kind of attack, however, is not what we address in this paper.

On the other hand, a compromised node can advertise forged routing information with a spoofed source address indicating that it came from other legitimate nodes (e.g. any of its neighbor nodes) to cause routing inconsistencies or to create routing loop. Since routing advertisements are locally broadcast messages (i.e. one-to-many), without an asymmetric mechanism, any compromised receiver could forge messages from the sender (i.e. impersonating other legitimate nodes). Even if groups of neighboring nodes establish a group key through the secure communication channels that were established between each pair of neighboring nodes as in our network assumption, the group key is also shared by a compromised node, seemingly no different from the neighboring legitimate nodes. Since the adversary can extract cryptographic keys, including the group key, from a captured node, establishing group key is useless for this problem. As a result, without an asymmetric mechanism, a compromised node could forge its identity to impersonate other legitimate nodes during routing advertisements, which leads routing inconsistencies.

3.3   ARMS' Goal

ARMS' main goal is the broadcast authentication of a sensor node's routing messages. While any published broadcast authentication protocols in sensor networks are not suitable for this goal and broadcast authentication of a sensor node's routing messages seem to have not been drawn much attention, it is the problem to be solved by all means for the routing security. Most of the routing protocols in sensor networks work by exchanging routing information among sensor nodes in distributed manner. In that situation, a single compromised node can influence a whole network topology by advertising malicious routing information [5]. Therefore, in the presence of several compromised nodes that inject inconsistent routing information, secure routing protocols must be robust by detecting and isolating them from legitimate nodes.

For designing such routing protocols, a mechanism for broadcast authentication of a sensor node's routing messages should be provided. Otherwise, by impersonating other legitimate nodes during routing advertisements, a compromised node might lay the blame upon other legitimate nodes to avoid being detected. As seen in Figure 1, a compromised node (node A) can exert a bad influence on routing topology without revealing its identity, even making the other legitimate node (node D) being suspected of advertising inconsistent routing information. Such situation makes the detection of a compromised node extremely hard or impossible. Thus, broadcast authentication of a sensor node's routing messages can be regarded as a fundamental mechanism for designing secure routing protocols.

Although ARMS can be used as a basis for the routing security, other types of local broadcast messages like an advertisement message containing meta-data in data-negotiation protocol [10] can also use ARMS for the authentication of a sensor node's broadcast messages. In this paper, however, we focus our description on the authentication of a sensor node's broadcast routing message.

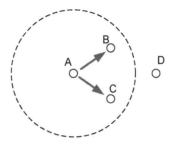

Fig. 1. A compromised node (node A) is advertising a forged routing message ,which says "I am node D , and I have an extremely low quality route to a base station", to its neighboring nodes (node B and C). Then, node B and C will not forward data through the legitimate node D.

3.4   Previously Proposed Schemes

ARMS requires only a small memory space of only 48 bytes for storing six 8-byte tokens at a broadcast sender side. That is the big difference between ARMS and the previous proposals for broadcast authentication [8][12], which require huge memory space for storing a one-way key chain. Since those proposals are designed for authentication of base station's broadcast messages, a resource-constrained sensor node does not have such memory space for storing that long one-way key chain.

In the previous schemes, the scheduled key disclosure delays authenticating broadcast messages, which can cause delayed propagation of routing information and lead to routing inconsistency. Differently from the previous schemes, ARMS authenticates a received broadcast message immediately. As a result, buffering of the received broadcast messages, which would wait until disclosure of a delayed MAC (Message Authentication Code) key in the previous schemes, is not needed. Immediate authentication is quite a good property for advertising routing messages in that the authentication delay could cause delayed propagation of routing information and lead to routing inconsistency. Delayed authentication and buffering also provide attackers with opportunities of denial-of-service (DoS) attacks. Moreover, ARMS does not require time synchronization between a broadcast sender and its neighboring receivers. Every sensor node broadcasts its routing messages, so every sensor node is a broadcast sender. It might be troublesome if each sensor node has to keep all its neighbors' time schedules and to operate on these schedules. In addition, ARMS has ability to mitigate the effect of packet loss.

4 ARMS: An Authenticated Routing Message in Sensor Networks

In this section, we present our scheme for the authentication of a sensor node's broadcast routing message. In following subsections, we introduce ARMS basic idea and the basic scheme first, and then describe the practical ARMS scheme in detail and ARMS phases.

4.1 ARMS Basic Idea

The basic idea of ARMS is that a broadcast sender makes consecutive broadcast packets uniquely related, and receivers authenticate broadcast packets immediately by verifying that relation. The relation is defined by a public one-way hash function F. A broadcast packet of the basic scheme consists of four elements as shown in Figure 2.

To broadcast a routing message, the broadcast sender picks a 64-bit random number K_N, which will be used as the one-time MAC key of a next broadcast packet, and configures its hash value as the first element of the current packet. To start broadcasting a routing message, as in μTESLA [12], a commitment is needed to set up neighboring nodes. Figure 3(a) and 3(b) show that a broadcast sender (node A) randomly picks 64-bit number K_1 as the one-time MAC key of the 1st packet and sets up its neighbor nodes by distributing the commitment $F(K_1)$. A commitment is the hash value of the one-time MAC key of the 1st broadcast packet from node A. Node A distributes a commitment to each neighbor node through secure communication channels by unicast. This process can be done during the establishment of secure communication channels with each neighbor node.

In Figure 3(b), to broadcast 1st routing message, node A picks a 64-bit random number K_2 as the one-time MAC key of the 2nd packet. Node A configures its hash value $F(K_2)$ and previously chosen number K_1 as the first and the second elements of the 1st packet, respectively. Since only the sender who generated the hash value $F(K_1)$ in the previous packet knows the current one-time MAC

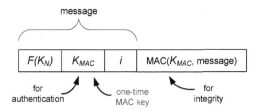

Fig. 2. The broadcast packet structure of the basic scheme. It consists of four elements, which are the hash value of the next broadcast message's MAC key ($F(K_N)$), the one-time MAC key of its own packet (K_{MAC}), routing information (i), and MAC (Message Authentication Code) of the packet, generated with the one-time MAC key ($\mathrm{MAC}(K_{MAC}, F(K_N)|K_{MAC}|i)$).

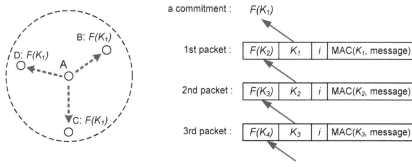

(a) A commitment distribution.

(b) The relation between packets in the basic scheme.

Fig. 3. In (b), each arrow indicates the relation defined by a public one-way hash function F

key K_1, none cannot make a legitimate broadcast packet except node A. Each neighbor node authenticates the 1st packet by hashing the second element (the one-time MAC key) of the 1st packet and comparing it with the commitment $F(K_1)$. Also, by generating MAC with the one-time MAC key of the packet, receivers are assured of data integrity. If being assured of data integrity, receivers then store the first element of the 1st packet $F(K_2)$ for authentication of the 2nd broadcast packet. If there is anything wrong either in authentication or in data integrity, the packet is discarded. Subsequent routing messages are generated and authenticated in the same manner as described above.

Consequently, the first element of the packet can be considered as a challenge for checking the next packet's authenticity and we can consider the second element of the packet as the response for the challenge. So, the second element of the packet is used for authentication and also considered as one-time MAC key.

Since a broadcast packet includes its own one-time MAC key, receivers can authenticate the packet immediately. Our concern is only the local broadcast which is a wireless one-hop transmission, not a multi-hop transmission. So, broadcast packet is received by all neighbors simultaneously. There is no case that any one of neighbors receives the broadcast packet earlier than other nodes. Even if there is extremely fine difference in receiving time, it is impossible for a malicious node to make a forged packet for that negligibly short moment, which requires the re-generation of MAC using the just received one-time MAC key.

In basic scheme, receivers store only one challenge per each sender, which is the hash value of the the one-time MAC key of the next broadcast packet. A broadcast sender stores only one response, which will be used as the one-time MAC key of the next broadcast packet.

4.2   Detailed ARMS Scheme

Due to its simplicity, the basic scheme described above seems suitable for a sensor node's routing advertisement messages. However, even a single packet loss breaks the relation between two consecutive broadcast packets and forces the broadcast sender to re-distribute a new commitment.

Here, we present our ARMS scheme mitigating the effect of packet loss, a practical version of the basic scheme. Figure 4 shows the relations among ARMS packets and the difference between ARMS and the basic scheme. Only the first (a challenge) and the second (a response) elements of the packet are presented in the figure, where each arrow indicates the relation defined by a public one-way hash function F. In ARMS, differently from the basic scheme, a set of four consecutive packets are uniquely related and these sets repeatedly appear in every other packets in a twisted pattern. The first packet of each set is called *a base*, so a base and a non-base packets alternate in sequence of packets. Once the packet is authenticated (and also assured of integrity), both the first and the second elements of the just authenticated packet are now treated as challenges for the following packets. So, receivers store both the first (named a challenge(L)) and the second (named a challenge(R)) elements of the most recently authenticated packet as challenges. Receivers authenticate the following packet by verifying relation between its second element (a response) and the challenges from the most recently authenticated packet.

As seen in Figure 4, while a single packet loss breaks the relation between the 3rd and the 5th packets in basic scheme, in ARMS the 3rd packets still can be authenticated with the challenges from the 1st packet. Also, even if two consecutive packets are lost, the 7th packet still can be authenticated with the challenges from the 4th packet. In this way, ARMS can tolerate packet loss unless two or more (when receivers' challenges are of a non-base packet) or three or more (when receiver's challenges are of a base packet) consecutive ARMS packets are lost.

ARMS packet has the same structure as the basic scheme's (Figure 2). The second element of the ARMS packet is used as the response for the challenges and also as a one-time MAC key of the packet, and the fourth element is MAC generated with the one-time MAC key of the packet.

A broadcast sender stores four tokens which will be used in the following ARMS packets. A broadcast sender also stores a current commitment pair which is the first and the second elements of the most recently broadcasted ARMS packet. A broadcast sender sends the current commitment pair to a newly discovered neighbor node through a secure unicast channel when a new neighbor is discovered. The current commitment pair will also be sent to the neighbor node that fails to tolerate packet loss. We explain the detailed usage of the current commitment pair later in next subsection. Figure 5 shows a concrete example of ARMS packets' relations and the sender's memory at the corresponding moment. In the figure, only the first and the second elements of the packets are presented.

A Sender side. A sender configures ARMS packet properly and broadcasts ARMS packets orderly so that a set of four consecutive packets are uniquely related and these sets repeatedly appear in every other packets in a twisted pattern.

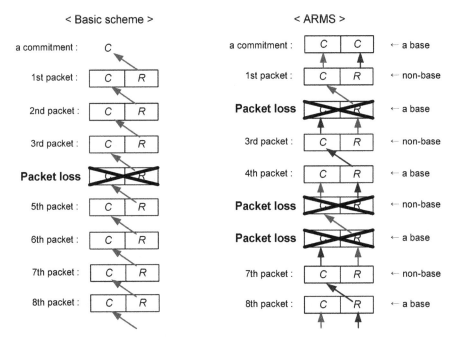

Fig. 4. The relations among ARMS packets. Only the first (a challenge) and the second (a response) elements of the packet are presented. Each arrow indicates the relation defined by a public one-way hash function F.

A sender picks a 64-bit random number (which will be used as the one-time MAC key of the last packet of a set of four uniquely related packets) at intervals of every other packet, more precisely at intervals of every non-base packet.

For example, to broadcast the 2nd packet (a non-base packet) in Figure 5, the sender configures $F(F(b))$ and a from memory as the first and the second elements of the 2nd packet respectively, and generates MAC with the one-time MAC key a. Then, the sender picks a 64-bit random number c and stores c and its hash value $F(c)$, with previously chosen number b and its hash value $F(b)$. And the sender stores $F(F(b))$ and a as the current commitment pair. To broadcast the 3rd packet (a base packet), the sender hashes $F(c)$ twice, stores $F(F(c))$ in memory, configures $F(F(F(c)))$ as the first element and $F(b)$ from memory as the second element of the 3rd packet, and also generates MAC with the one-time MAC key $F(b)$. And the sender stores $F(F(F(c)))$ and $F(b)$ as the current commitment pair. To broadcast the 4th packet (a non-base), the sender configures $F(F(c))$ and b from memory as the first and the second elements respectively, and generates MAC with the one-time MAC key b. Then, the sender picks another 64-bit random number d and stores d and its hash value $F(d)$, with previously chosen number c and its hash value $F(c)$. And the sender stores $F(F(c))$ and b as the current commitment pair. Subsequent procedures for the sender are the same as we described above. In this way, the sender can broadcast uniquely related ARMS packets.

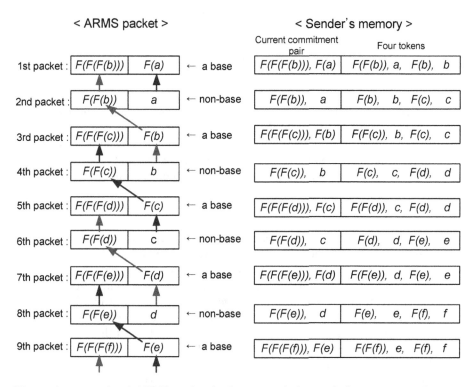

Fig. 5. An example of ARMS packets' relations and the sender's memory at the corresponding moment. Only the first and the second elements are presented. Each arrow indicates the relation defined by a public one-way hash function F. A broadcast sender stores four tokens which will be used in the following ARMS packets and stores also a current commitment pair which is the first and the second elements of the most recently broadcasted ARMS packet.

A Receiver side. A receiver stores the challenges (a challenge(L) and a challenge(R)) and knows whether the challenges are of the base packet. Keeping those information, a receiver authenticates the ARMS packets immediately and tolerates packet loss.

As an example, we assume that the receiver has just authenticated the 4th packet from the sender node A in Figure 5. Since the packet is authenticated, the first and the second elements of the packet are stored as the challenges in the receiver's neighbor table, as shown in Figure 6. Because the 4th packet is a non-base packet, the flag B, which indicates whether the challenges are of the base packet, is set to false. The receiver's authentication procedures may differ depending on the value of B. The initial value is distributed along with the initial commitment pair by the broadcast sender. We explain how the receiver authenticates the broadcast packets depending on B and how it keeps track of the value. We assume that the broadcast packet is just received from a node claiming A.

Neighbor ID	Cost	challenges		B
		(L)	(R)	
A	i	$F(F(c))$	b	false
...	...	...	...	...

Fig. 6. The neighbor table of the receiver. The receiver has just authenticated the 4th packet from the sender node A in Figure 5. Routing information from the authenticated ARMS packet is stored in Cost field, the first and the second elements of the most recently authenticated packet are stored in challenge(L) and challenge(R) fields respectively, and B field contains a binary value, which indicates whether the challenges are of the base packet.

(a) B = false (e.g. the receiver's challenges are of the 4th packet in Figure 5): The receiver first hashes the second element (a response) of the just received packet and compares it with the challenge(L) from the neighbor table. If it is true and integrity is also verified, the first and the second elements of the packet are stored as the new challenges. Since a base and non-base packet alternate, this packet must be a base packet (e.g. the 5th packet in Figure 5), so the flag B is set to true. Otherwise, the receiver hashes that hashed value again and compares it with the challenge(L) from the neighbor table to see if the packet has been lost. If it is true and integrity is also verified, the first and the second elements of the packet are stored as the new challenges. Since the previous packet was lost in this case and this packet must be a non-base packet (e.g. the 6th packet in Figure 5), B is set to false. Otherwise, the packet is discarded. Thus, when B is false, the receiver hashes the response of the just received packet twice at worst case and tolerates packet loss unless two or more consecutive packets are lost.

(b) B = true (e.g. the receiver's challenges are of the 5th packet in Figure 5): The receiver first hashes the second element (a response) of the just received packet and compares it with the challenge(R) from the neighbor table. If it is true and integrity is also verified, the first and the second elements of the packet are stored as the new challenges. Since a base and non-base packet alternate, this packet must be a non-base packet (e.g. the 6th packet in Figure 5), so the flag B is set to false. Otherwise, the receiver hashes that hashed value again and compares it with the challenge(L) from the neighbor table to see if the packet has been lost. If it is true and integrity is also verified, the first and the second elements of the packet are stored as the new challenges. Since the previous packet was lost in this case and this packet must be a base packet (e.g. the 7th packet in Figure 5), B is set to true. Otherwise, the receiver hashes that hashed value again and compares it with the challenge(L) from the neighbor table to see if two consecutive packet have been lost. If it is true and integrity is also verified, the first and the second elements of the packet are stored as the new challenges. Since the previous two consecutive packets were lost in this case and this packet must be a non-base packet (e.g. the 8th packet in Figure 5), B is set to false. Otherwise, the packet is discarded. Thus, when B is true, the receiver hashes the response of the just received packet three times at worst case and tolerates packet loss unless three or more consecutive packets are lost.

4.3   ARMS Phases

ARMS has four phases: Initial setup, Receiver setting, Broadcast ARMS packets, and Immediate authentication by receivers, as shown in Figure 7.

Initial setup. A broadcast sender picks two 64-bit random numbers, prepares the four tokens for the following ARMS packets and makes an initial commitment pair before the establishment of secure communication channels among neighbor nodes. For example, $F(F(F(b)))$ and $F(a)$ of the 1st ARMS packet in Figure 5 can be considered as an initial commitment pair and also $F(F(b))$, a, $F(b)$ and b as the four tokens that should be stored in memory.

Receiver setting. When the broadcast sender discovers neighbors and establishes secure communication channels, the sender distributes the initial commitment pair (which was made during initial setup) and the flag B associated with the initial commitment pair through secure communication channels by unicast.

Also, when the broadcast sender receives a Receiver Setting Request (RSR) message from any of its neighbors, the broadcast sender unicasts the current commitment pair and the flag B associated with the current commitment pair to that neighbor node through a secure communication channel. We explain the usage of the Receiver Setting Request (RSR) later in *Immediate authentication by receivers* part.

Broadcast ARMS packets. The sender broadcasts ARMS packets as we described in the previous subsection.

Immediate authentication by receivers. By initial receiver setting phase, the receiver receives the initial commitment pair and the flag B from the sender through a secure unicast communication channel. This initial commitment pair is used as the receiver's initial challenges. Using those information, a receiver authenticates the ARMS packet immediately and tolerates packet loss. Updating and keeping those information, subsequent ARMS packets are authenticated by receivers.

However, either if two or more consecutive packets are lost (when the receiver's challenges are of the non-base packet) or if three or more consecutive packets are lost (when the receiver's challenges are of the base packet), ARMS fails to tolerate the effect of the packet loss. Since ARMS is routing message, we can set up a Maximum Time Interval (MTI) between two consecutive ARMS packets. When the flag B is true (which means the receiver can tolerate two consecutive ARMS packet loss), if no broadcast packet is authenticated from a particular neighbor for $3 \times MTI$, it is considered as three consecutive ARMS packet loss and receiver setting is needed again. And, when the flag B is false, (which means the receiver can tolerate one ARMS packet loss), if no broadcast packet is authenticated from a particular neighbor for $2 \times MTI$, it is considered as two consecutive ARMS packet loss and receiver setting is needed again. Receiver setting is done by sending a Receiver Setting Request (RSR) message to the corresponding sender through a secure unicast communication channel.

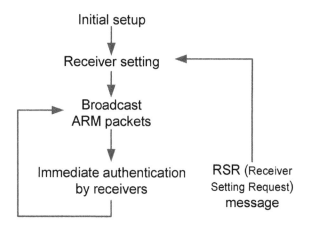

Fig. 7. ARMS phases

5   Evaluation

In this section, we evaluate the packet size of ARMS and the effect of packet loss rate on the generation of RSR (Receiver Setting Request) messages.

As seen in Figure 8, the first and the second elements of the ARMS packet occupy 8 bytes each. Although conventional size of MAC (Message Authentication Code) is 8 bytes or 16 bytes, a 4-byte MAC is found to be not detrimental in sensor networks by Karlof *et al.* [6]. Hence, we employ a 4-byte MAC, the last element of the ARMS packet. Receivers authenticate ARMS packet only through an 8-byte response (the second element of the ARMS packet). A 4-byte MAC in ARMS is only used for integrity check only if an 8-byte response is verified, so MAC in ARMS has something to do with wireless channel error, not an adversary's attack. This can be considered as a double lock whose outer lock is much stronger than an inner lock, so adversary still has a 1 in 2^{64} chance of guessing the right response. Consequently, ARMS has a total of 20 extra bytes added to routing information. Considering 30-byte packets in sensor networks, the remaining space is quite sufficient for routing information.

Also, the computations performed in sender and receiver side are just one-way hash functions which can be done with a block cipher like RC5. Thus, this amount of work is easily performed.

Figure 9 shows the impact of packet loss rate on the frequency of the generation of RSR messages. As packet loss rate increases, a receiver tends to generate more RSR messages. If the packet loss rate reaches over 90%, a receiver sends RSR messages approximately at intervals of 2.5 MTI (Maximum Time Interval)

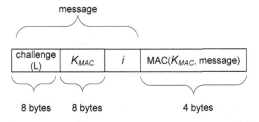

Fig. 8. The byte size of each element in ARMS packet. The first and the second elements of the packet hold 8 bytes each. The size of the third element (routing information) depends on a routing protocol being used. The size of the last element (MAC) is 4 bytes.

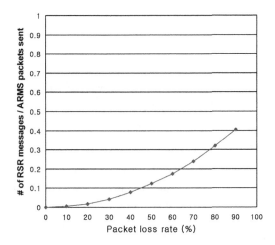

Fig. 9. The impact of packet loss rate on the frequency of the generation of RSR messages

on the average, which result in about 40% RSR messages over the ARMS packets sent by a sender. Where packet loss rate is less than 50% , only a small number of RSR messages are generated, which is considered as negligible overhead. However, as packet loss rate increases over 50%, RSR messages are generated too often, which incurs a large amount of communication overhead. This is because ARMS is designed for mitigating the effect of packet loss over reasonably lossy wireless channel.

6 Conclusion

In this paper, we have presented ARMS to address the broadcast authentication of a sensor node's routing message. Differently from broadcast authentication of a base station's message, due to severe hardware constraints, a sensor node cannot afford the long one-way key chain. Moreover, the delayed authentication feature in previous schemes seems undesirable for routing message propagation. ARMS requires only 48-byte memory space in sender side and 16 bytes in corresponding receivers' side. Since ARMS authenticates broadcast messages immediately and does not require time synchronization among sensor nodes, ARMS is quite suitable for broadcast authentication of a sensor node's routing message. Also, under reasonable circumstances (less than 50% packet loss rate), ARMS mitigates the effect of packet loss with negligible communication overhead. Armed with ARMS, we believe researchers will be able to design a secure routing protocol that detects and isolates compromised nodes.

Acknowledgement

This work was supported by 2004 Hongik University Research Fund.

References

1. Haowen Chan, Adrian Perrig, and Dawn Song. Random key predistribution schemes for sensor networks. In IEEE Symposium on Security and Privacy, May 2003.
2. Laurent Eschenauer and Virgil D. Gligor. A key-management scheme for distributed sensor networks. In 9th ACM Conference on Computer and Communication Security (CCS), November 2002.
3. Carl Hartung, James Balasalle, Richard Han, Node Compromise in Sensor Networks: The Need for Secure Systems, Technical Report CU-CS-990-05, Department of Computer Science University of Colorado at Boulder, January 2005
4. W.R. Heinzelman, J. Kulik, H. Balakrishnan, Adaptive protocols for information dissemination in wireless sensor networks, Proceedings of the ACM MobiCom99, Seattle, Washington, 1999, pp. 174.185.
5. Chris Karlof and David Wagner, Secure Routing in Wireless Sensor Networks: Attacks and Countermeasures, to appear First IEEE InternationalWorkshop on Sensor Network Protocols and Applications, May 2003
6. Chris Karlof, Naveen Sastry, and David Wagner TinySec: A Link Layer Security Architecture for Wireless Sensor Networks. Proceedings of the Second ACM Conference on Embedded Networked Sensor Systems (SensSys 2004), November 2004.
7. Donggang Liu and Peng Ning. Establishing pairwise keys in distributed sensor networks. In 10th ACM Conference on Computer and Communications Security (CCS), October 2003.
8. Donggang Liu, Peng Ning, Multi-Level μ-TESLA: A Broadcast Authentication System for Distributed Sensor Networks, Submitted for journal publication. Also available as Technical Report, TR-2003-08, North Carolina State University, Department of Computer Science, March 2003.
9. Adrian Perrig, Ran Canetti, J.D. Tygar, and Dawn Song. Efficient authentication and signing of multicast streams over lossy channels. In IEEE Symposium on Security and Privacy, May 2000.
10. Adrian Perrig, Ran Canetti, Dawn Song, and J. D. Tygar. Efficient and secure source authentication for multicast. In Network and Distributed System Security Symposium, NDSS 01, February 2001.

11. Adrian Perrig, John Stankovic, and David Wagner Security in wireless sensor networks. Communications of the ACM, 47(6), June 2004, Special Issue on Wireless sensor networks, pp.53- 57.
12. Adrian Perrig, Robert Szewczyk, Victor Wen, David Culler, and J.D. Tygar. SPINS: Security protocols for sensor networks. In The Seventh Annual International Conference on Mobile Computing and Networking (MobiCom 2001), 2001.
13. Anthony D.Wood, John A. Stankovic. Denial of Service in Sensor Networks. IEEE Computer, 35(10):54-62, 2002.

Security Analysis and Improvement of Return Routability Protocol

Ying Qiu[1], Jianying Zhou[1], and Robert Deng[2]

[1] Institute for Infocomm Research (I^{2}R) 21, Heng Mui Keng Terrace,
Singapore, 119613
{qiuying, jyzhou}@i2r.a-star.edu.sg
[2] Singapore Management University (SMU) 80 Stamford Road, Singapore 178902
robertdeng@smu.edu.sg

Abstract. Mobile communication plays a more and more important role in computer networks. How to authenticate a new connecting address belonging to a said mobile node is one of the key issues in mobile networks. This paper analyzes the Return Routability (RR) protocol and proposes an improved security solution for the RR protocol without changing its architecture. With the improvement, three types of redirect attacks can be prevented.

Keywords: Authentication, Redirect Attacks, Security, MIPv6.

1   Introduction

Mobile networking technologies, along with the proliferation of numerous portable and wireless devices, promise to change people's perceptions of the Internet. In true mobile networking, communications activities are not disrupted when a user changes his/her device's point of attachment to the Internet - all the network reconnections occur automatically and transparently to the user. The IETF RFC 3775 [1] supports mobile networking by allowing a mobile node to be addressed by two IP addresses, a home address and a care-of address. The former is an IP address assigned to the mobile node within its subnet prefix on its home subnet and the latter is a temporary address acquired by the mobile node while visiting a foreign subnet. The dual address mechanism in Mobile IP network allows packets to be routed to the mobile node regardless of its current point of attachment and the movement of the mobile node away from its home subnet is transparent to transport and higher-layer protocols. Fig 1 shows the basic operation in mobile IPv6.

One of the major features in Mobile IPv6 is the support for "Route Optimization" as a built-in fundamental part of the Mobile IPv6 protocol. The integration of route optimization functionality allows direct routing from any correspondent node (CN) to any mobile node (MN), without needing to pass through the mobile node's home sub-net and be forwarded by its home agent (HA), and thus eliminates the problem of "triangle routing". Route optimization in Mobile IPv6 requires that the MN, HA and the CNs maintain a Binding Cache. A binding

M. Burmester and A. Yasinsac (Eds.): MADNES 2005, LNCS 4074, pp. 174–181, 2006.
© Springer-Verlag Berlin Heidelberg 2006

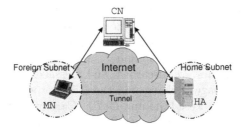

Fig. 1. Basic operation in mobile IPv6

is the association of a MN's home address (HoA) with a care-of address (CoA) for that mobile node, along with the remaining lifetime of that association. A mobile node uses Binding Update (BU) messages to notify its CNs or its HA of its current binding. Unfortunately, unauthenticated binding update messages provide intruders with an easy means to launch "Redirect Attacks", i.e., malicious acts which redirect traffic from the correspondent nodes to destinations chosen by intruders. Therefore, security of the binding update messages is of para-mount importance for Mobile IPv6 to meet its basic security requirements.

In IETF RFC 3775 [1], the Return Routability protocol (RR) is deployed to secure binding updates from MN to CNs. The basic RR mechanism consists of two checks, a home address check and a care-of-address check.

In the paper, we will analyze the RR mechanism and point out three attacks to the RR protocol, and finally propose a solution without changing the RR architecture.

The notations used throughout this paper are listed below:

h( )   a one-way hash function, such as SHA1 [2].

prf(k, m)   a keyed pseudo random function - often a keyed hash function [3]. It accepts a secret key k and a message m, and generates a pseudo random output. This function is used for both message authentication and cryptographic key derivations.

e(k ,m)   encryption of message m with a secret key k.

m|n   concatenation of two messages m and n.

MN   mobile node HA home agent of a mobile node.

CN   correspondent node of a mobile node.

CNA   IP address of CN.

HoA   home address of a mobile node.

CoA MN's care-of address when it visits a foreign network.

2   Brief Review of RR Protocol

In RFC 3775's Return Routability (RR) protocol [1], a CN keeps a secret key k_{CN} and generates a nonce at regular intervals, say every few minutes. CN uses the same key k_{CN} and nonce with all the mobile nodes it is in communication with, so that it does not need to generate and store a new nonce when a new mobile node contacts it. Each nonce is identified by a nonce index. When a new nonce is generated, it must be associated with a new nonce index, e.g., j. CN keeps both the current value of N_j and a small set of previous nonce values, $N_{j-1}, N_{j-2},$. Older values are discarded, and messages using them will be rejected as replays. Message exchanges in the RR protocol are shown in Fig 2, where the $HoTI$ (Home Test Init) and $CoTI$ (Care-of Test Init) messages are sent to CN by a mobile node MN simultaneously. The HoT (Home Test) and CoT (Care-of Test) are replies from CN. All RR protocol messages are sent as IPv6 "Mobility Header" in IPv6 packets. In the representation of a protocol message, we will use the first two fields to denote source IP address and destination IP address, respectively. We will use CNA to denote the IP address of the correspondent node CN.

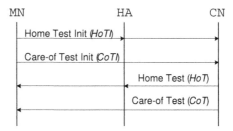

Fig. 2. Return Routability protocol

When MN wants to perform route optimization, it sends

$$HoTI = \{HoA, CNA, r_H\}$$

and

$$CoTI = \{CoA, CNA, r_C\}$$

to CN, where r_H and r_C are random values used to match responses with requests. $HoTI$ tells MN's home address HoA to CN. It is reverse tunneled through the home agent HA, while $CoTI$ informs MN's care-of address CoA and is sent directly to CN.

When CN receives $HoTI$, it takes the source IP address of $HoTI$ as input and generates a home keygen token

$$KT_H = prf(k_{CN}, HoA|N_j|0)$$

and replies MN with

$$HoT = \{CNA, HoA, r_H, KT_H, j\},$$

where | denotes concatenation and the final "0" inside the pseudo random function is a single zero octet, used to distinguish home and care-of cookies from each other. The index j is carried along to allow CN later efficiently finding the nonce value N_j that it used in creating the token KT_H. Similarly, when CN receives $CoTI$, it takes the source IP address of $CoTI$ as input and generates a care-of keygen token

$$KT_C = prf(k_{CN}, CoA|N_i|1)$$

and sends

$$CoT = \{CNA, CoA, r_C, KT_C, i\}$$

to MN, where the final "1" inside the pseudo random function is a single octet "0x01". Note that HoT is sent via MN's home agent HA while CoT is delivered directly to MN.

When MN receives both HoT and CoT, it hashes together the two tokens to form a session key

$$k_{BU} = h(KT_H|KT_C),$$

which is then used to authenticate the correspondent binding update message to CN:

$$BU = \{CoA, CNA, HoA, Seq\#, i, j, MAC_{BU}\},$$

where $Seq\#$ is a sequence number used to detect replay attack and

$$MAC_{BU} = prf(k_{BU}, CoA|CNA|HoA|Seq\#|i|j)$$

is a message authentication code (MAC) protected by the session key k_{BU}. MAC_{BU} is used to ensure that BU was sent by the same node which received both HoT and CoT. The message BU contains j and i, so that CN knows which nonce values N_j and N_i to use to first recompute KT_H and KT_C and then the session key k_{BU}. Note that CN is stateless until it receives BU and verifies MAC. If MAC is verified positive, CN may reply with a binding acknowledgement message

$$BA = \{CNA, CoA, HoA, Seq\#, MAC_{BA}\},$$

where $Seq\#$ is copied from the BU_{CN} message and

$$MAC_{BA} = prf(k_{BU}, CNA|CoA|HoA|Seq\#)$$

is a MAC generated using k_{BU} to authenticate the BA message. CN then creates a binding cache entry for the mobile node MN. The binding cache entry binds HoA with CoA which allows future packets to MN be sent to CoA directly.

An example implementation of the binding cache at CN is shown in Fig 3, where HoA is used as an index for searching the binding cache, the sequence number $Seq\#$ is used by CN to check the freshness of binding updates. Each binding update sent by MN must use a $Seq\#$ greater than (modulo 2^{16}) the one sent in the previous binding update with the same HoA. It is not required, however, that the sequence number value strictly increase by 1 with each new binding update sent or received [1]. Note that the session key k_{BU} is not kept in the cache entry. When CN receives a binding update message, based on the

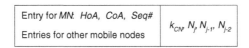

| Entry for *MN*: *HoA*, *CoA*, *Seq#* | $k_{CN}, N_j, N_{j-1}, N_{j-2}$ |
| Entries for other mobile nodes | |

Fig. 3. A binding cache implementation at CN in the RR protocol

nonce indexes i and j in the message, it recomputes the session key using k_{CN} and the list of the most recent nonce values, say $\{N_j, N_{j-1}, N_{j-2}\}$, and then verifies BU using the newly computed session key.

The mobile node MN maintains a Binding Update List for each binding update message sent by it, for which the lifetime has not yet expired. A binding update list for a correspondent node CN consists of CNA, MN's HoA and CoA, the remaining lifetime of the binding, the maximum value of the sequence number sent in previous binding updates to CN and the session key k_{BU}.

3   Redirect Attacks to RR Protocol

Obviously, the RR protocol protects binding updates against intruders who are unable to monitor the HA-CN path and the MN-CN path simultaneously. However, one has no reason to assume that an intruder will monitor one link and not the other, especially when the intruder knows that monitoring a given link is particularly effective to expedite its attack. Even worse, we demonstrate that the RR protocol can be attacked under its original assumption of no simultaneous monitor of both the HA-CN path and the MN-CN path.

3.1   Session Hijacking Attacks

Let's consider the scenarios showed in Fig 4, a mobile node MN_1 is communicating with a correspondent node CN. An intruder sends a forged binding update message (or replays an old binding update message) to CN, claiming that MN_1 has moved to a new care-of-address belonging to a node MN_2. If CN accepts the fake binding update, it will redirect to MN_2 all packets that are intended to MN_1. This attack allows the intruder to hijack ongoing connections between MN_1 and CN or start new connections with CN pretending to be MN_1. This is an "outsider" attack since the intruder tries to redirect other nodes' traffic. Such an attack may result in information leakage, impersonation of the mobile node MN_1 or flooding of MN_2.

This attack is serious because MN_1, MN_2, CN and the intruder can be any nodes anywhere on the Internet. All the intruder needs to know is the IP addresses of MN_1 and CN. Since there is no structural difference between a mobile node home address and a stationary IP address, the attack works as well against stationary Internet nodes as against mobile nodes. The deployment of a binding update protocol without security could result in breakdown of the entire Internet [4].

In the case of the static IPv6 without mobility (which is equivalent to the mobile node MN at its home subnet in Mobile IPv6), to succeed in the attack, the intruder must be constantly present on the CN-HA path. In order to redirect CN's traffic intended for MN to a malicious node, the intruder most likely has to get control of a router or a switch along the CN-HA path. Furthermore, after taking over the session from MN, if the malicious node wants to continue the session with CN while pretending to be MN, the malicious node and the router need to collaborate throughout the session. For example, the router tunnels CN's traffic to the malicious node and vise versa.

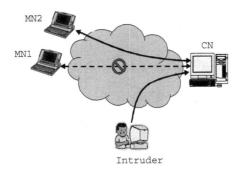

Fig. 4. Session hijacking attacks

In the case of Mobile IPv6, the effort committed to break the RR protocol to launch a session hijacking attack could be considerably lesser. Assume that MN_1 and CN are having an on-going communication session and the intruder wants to redirect CN's traffic to his collaborator MN_2. The intruder monitors the CN-HA path (i.e., anywhere from MN_1's home network to CN's network) to obtain HoT, extracts the home keygen token KT_H and sends it to MN_2.

Upon receiving KT_H, MN_2 sends a $CoTI$ to CN and CN will reply with a care-of keygen token KT_C. MN_2 simply hashes the two keygen tokens to obtain a valid binding key, and uses the key to send a binding update message to CN on behalf of MN_1. The binding update will be accepted by CN which will in turn direct its traffic to MN_2.

3.2 Movement Halting Attacks

Another related attack is when a mobile node MN rapidly moves from one care-of ad-dress CoA to another CoA'. Since MN runs the RR protocol whenever it moves to a new location, an intruder can intercept the care-of keygen token KT_C in the current RR session and the home keygen token KT_H in the next RR session, hash the two keygen tokens to get a valid binding key, and then send a binding update message with the CoA in the current session to the correspondent node. The correspondent node will still send its traffic back to CoA. Hence, MN, which has moved to CoA', will not receive data from the correspondent node. Note that in this attack the attacker does not have to intercept the two keygen tokens at the "same time".

3.3   Traffic Permutation Attacks

The RR protocol is also subject to a "traffic permutation" attack. Consider a correspondent node which provides on-line services to many mobile clients (Fig 5). An intruder can simply eavesdrop on the RR protocol messages to collect keygen tokens on the border between the correspondent node and the Internet. The intruder then hashes random pairs of keygen tokens to form binding keys, and sends binding update messages to the correspondent node.

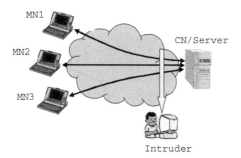

Fig. 5. Intruder attacks an on-line server

Such a forged binding update message will be accepted by the correspondent node with probability 1/4. This will cause redirection of traffic to randomly selected mobile clients and eventually bring down the services of the correspondent node.

4   Improvement of RR Protocol

The attacks outlined in the above section are due to the decoupling of HoA and CoA in RR messages. In the original RR protocol, the home keygen token

$$KT_H = prf(k_{CN}, HoA|N_j|0)$$

and the care-of keygen token

$$KT_C = prf(k_{CN}, CoA|N_i|1)$$

are delivered without any stated relationship. Any pair of home keygen token and care-of keygen token can generate a valid binding key

$$k_{BU} = h(KT_H|KT_C)$$

as long as the indexes, i and j, are still valid.

However, the attacks described in the above section can be prevented by modifying the RR protocol to include both CoA and HoA in the generation of home keygen token and care-of keygen token, respectively. In the improved RR protocol, HoA and CoA are bound together. (The modified parts are underscored.) A mobile node sends

$$HoTI' = \{HoA, CNA, CoA, r_H\}$$

and

$$CoTI' = \{CoA, CNA, HoA, r_C\}$$

to a CN, which replies with the home keygen token

$$KT'_H = prf(k_{CN}, HoA|N_j|CoA|0)$$

and the care-of keygen token

$$KT'_C = prf(k_{CN}, CoA|N_i|HoA|1).$$

Then the new binding key

$$k'_{BU} = h(KT'_H|KT'_C)$$

is valid only for the pair of HoA and its claimed CoA. Therefore the misuse of keygen tokens can be avoided.

5   Conclusion

In this paper, we first reviewed the Return Routability protocol in RFC 3775, then demonstrated three redirect attacks: Session Hijacking Attacks, Movement Halting Attacks and Traffic Permutation Attacks. We further pointed out that the attacks are due to the decoupling of HoA and CoA in RR messages. We also proposed an improved solution that provides much stronger security than the original RR protocol without changing its architecture.

References

1. D. Johnson, C. Perkins, and J. Arkko, "Mobility Support in IPv6", IETF RFC 3775, June 2004.
2. NIST, "Secure Hash Standard", NIST FIPS PUB 180, May 1993.
3. H. Krawczyk, M. Bellare, and R. Canetti, "HMAC: Keyed-Hashing for Messaging Authentication", IETF RFC 2104, February 1997.
4. T. Aura, "Mobile IPv6 Security", Proceedings of the 10th International Workshop on Security Protocols, LNCS 2467, Cambridge, UK, April 2002.

Dark Encounter Computations
(and Musings on Biological Networks)
(Extended Abstract)

René Peralta

Information Technology Laboratory, National Institute of Standards and Technology
rene.peralta@nist.gov

1 Introduction

Two models of distributed computation are described in which the agents are anonymous finite-state sensors interacting through a communication network whose dynamics (in case the agents are mobile) and/or topology they do not control or even know about. These models were recently introduced in a series of papers by Angluin et al. [3,4,2,5].

In one model, the agents are stationary but form a network of unknown topology. Furthermore, the pair-wise communication links in the network become available in a random and unpredictable infinite sequence. The sensors would like to compute a function of their inputs. While doing so they are, by design, or inherent limitation, not endowed with unique identities (hence the term "dark encounter computations" in the title of this note). Angluin et al. developed a theory about such networks that, among other results, showed that their computational power is dependent on topology. Perhaps contrary to expectation, the power is inversely proportional to the connectivity, with a straight-line configuration being the most powerful and the complete graph configuration being the weakest. This observation leads naturally to a number of questions related to the structure of the network. What topological properties of the network can be computed? (e.g., is the network a ring?). What properties can individual sensor nodes compute? (e.g., am I in a cycle?). These and related questions are answered in the affirmative in [2].

In the second model, agents are randomly mobile and communicate via pairwise interactions. As with the first model, the agents are anonymous and have no control of the sequence of interactions. A stochastic assumption ensures that every pair of agents eventually encounter each other.[1] Angluin et al. show that this model of computation turns out to be surprisingly powerful. In a type of computation called "population protocol" the networks are able to compute predicates such as parity and majority. Population protocols can also perform modular arithmetic (for details, see [4]).

There are, of course, a number of possible definitions of what it means for these networks to "compute" something. In the work of Angluin et al. the network is

---

[1] An example of such a network is RFID devices installed in cars. In this setup, sensors get to exchange information when two cars get sufficiently close to each other.

M. Burmester and A. Yasinsac (Eds.): MADNES 2005, LNCS 4074, pp. 182–185, 2006.

said to *stably compute* a function if all sensors eventually converge to the correct value even when their inputs are allowed to vary an unbounded but finite number of times.

It is natural to wonder how contrived this paradigm is. The question arises because most of these problems become uninteresting if the finite-state and/or the anonymity constraints are removed. In particular, if the sensors are endowed with enough memory, the nodes can assign each other unique identifiers and are then able to simulate a Turing Machine. Anonymity, however, is a plausible constraint in a variety of scenarios. Some of these are listed below. Each is deserving of a lengthy discussion. In this short note, however, I only elaborate on one of them.

- Biological networks (see discussion below);
- Nanotechnology: molecular-sized engineered devices for distributed computation (in a patient's bloodstream, for example) may well have extremely limited computational and (usable) memory resources;
- Limited memory use may be a desirable security property. For example, a captured spying device would ideally "know" very little about what it is doing, where it has been, and what it has seen. In this way, the purpose of the "mission" may remain secret;
- Even when network agents, such as RFID devices, have enough memory to hold unique identifiers, privacy concerns have been raised which may be addressed by designing the devices to be anonymous. This issue is being debated in the context of RFID chips on vehicles (see, for example, the controversy surrounding Texas bill HB2893 http://www.atsnn.com/story/132110.html).

2   Biological Networks

In biological networks of lower organisms, and despite the fact that the state space of member agents is typically huge (much larger than the number of agents), we can reasonably expect anonymous communication as a result of evolutionary processes. In their random evolutionary walk through the space of possible configurations and behaviors, it seems highly unlikely that lower organisms such as bacteria would develop the ability to call each other by name. Yet bacteria have been shown to exhibit a remarkable range of social behaviors. For example, they are able to coordinate sporulation in such a way that the collective behavior only occurs if enough of the members "think" it is a good idea. It is unclear to what extent the different known communication mechanisms of bacteria can be simulated (in a way that does not do away with fundamental properties of biological systems) by the networks of Angluin et al. The type of distributed problem-solving we are interested in here has been shown to occur via chemotactic signaling and quorum-sensing. However, these communication mechanisms do not, strictly speaking, involve pairwise interaction. Another communication mechanism, plasmid exchange, does involve pairwise interaction. But I am not aware of this mechanism being used in coordinating real-time bacterial behavior. Nevertheless, distributed computation via pair-wise communication has been documented in neurons (quite surprisingly, as most studied neuronal communication involves broadcasting to all neighboring neurons) [8]. This gives reason to suspect the phenomenon may not be uncommon in biological networks of lower organisms (it is clearly an important component of the social behavior of higher organisms such as humans). In my own observations of pond life through a microscope, I have observed many "networks" of algae which clearly coordinate behavior and communicate via pairwise interactions. I have no expertise, however, to investigate the question of whether the two observed phenomena (coordination and communication through pairwise interaction) are related.

Emerging areas of research aim at describing the fundamental properties of biological networks. Among the most prolific of these efforts are those of Barabasi's group (see http://www.nd.edu/~alb/). Here are, without elaboration, some morsels of wisdom from this body of work:

- the World Wide Web exhibits a topology which is not what one would expect. A computer scientist might expect it to be similar to Erdös's random graphs. A social scientist would expect it to be similar to "small-world" networks. As shown in [1], it is neither, and apparently the differences have important implications (with hindsight, one quickly realizes that random graphs in the sense of Erdös and Rényi [6] are not likely to be endowed with life-sustaining properties such as Maturana and Varela's autopoiesis [2]).
- the same structural properties that make the World Wide Web and the Internet resilient to local failures also makes them vulnerable to attack.

A different methodological approach to understanding biological networks seems implicit in Harvard's Bauer Center writings (see http://cgr.harvard.edu/research/biological.html). Part of its research mission statement reads as follows

"What subset of the space of possible networks do biological systems occupy? Can we explain why they occupy this subset? Possible answers to the last question include recycling of historical accidents, strong evolutionary constraints from the combined requirement for robustness and adaptability, and the requirement that any functional module can be constructed by a series of incremental improvements. "

In a recent report by the National Research Council's Committee on Network Science for Future Army Applications (executive summary is available at http://darwin.nap.edu/execsumm_pdf/11516.pdf), the Committee concludes that abstract networks are poorly understood and that such understanding is necessary to address the problem of securing computer networks. The report goes on to recommend that a new field, "network science", be funded.

Although the above discussion may seem far removed from network security issues, I contend that it is not. Autopoiesis, for example, views as a defining

---

[2] For a review of autopoiesis see, for example, [7].

characteristic of life that of being able to maintain structure (and hence functionality) while interacting with an environment that is constantly damaging, or outright destroying, components of said structure.

Acknowledgements

I am indebted to Kevin Mills for many helpful suggestions on this manuscript.

References

1. R. Albert, A. Barabasi, and H. Jeong. Scale-free characteristics of random networks: The topology of the world wide web. *Physica A: Statistical Mechanics and its Applications*, 281:69–77, June 2000.
2. Dana Angluin, James Aspnes, Melody Chan, Michael J. Fischer, Hong Jiang, and René Peralta. Stably computable properties of network graphs. In Viktor K. Prasanna, Sitharama Iyengar, Paul Spirakis, and Matt Welsh, editors, *Distributed Computing in Sensor Systems: First IEEE International Conference, DCOSS 2005, Marina del Rey, CA, USE, June/July, 2005, Proceedings*, volume 3560 of *Lecture Notes in Computer Science*, pages 63–74. Springer-Verlag, June 2005.
3. Dana Angluin, James Aspnes, Zoë Diamadi, Michael J. Fischer, and René Peralta. Urn automata. Technical Report YALEU/DCS/TR–1280, Yale University Department of Computer Science, November 2003.
4. Dana Angluin, James Aspnes, Zoë Diamadi, Michael J. Fischer, and René Peralta. Computation in networks of passively mobile finite-state sensors. In *PODC '04: Proceedings of the twenty-third annual ACM symposium on Principles of distributed computing*, pages 290–299. ACM Press, 2004.
5. Dana Angluin, James Aspnes, Michael J. Fischer, and Hong Jiang. Self-stabilizing population protocols. In *Ninth International Conference on Principles of Distributed Systems (pre-proceedings)*, pages 79–90, December 2005.
6. Erdös and Rényi. On the evolution of random graphs. *Publ. Math. Inst. Hung. Acad. Sci.*, 17(5), 1960.
7. Pier Luigi Luisi. Autopoiesis: a review and a reappraisal. *Naturwissenschaften*, 90:49–59, Feb 2003.
8. Elad Schneidman, Michael J. Berry, Ronen Segev, and William Bialek. Weak pairwise correlations imply strongly correlated network states in a neural population. *Nature*, April 2006.

Panel: Authentication in Constrained Environments

Panelists: Mike Burmester[1], Virgil Gligor[2], Evangelos Kranakis[3], Doug Tygar[4] and Yuliang Zheng[5]

Transcriber: Breno de Medeiros[1]

[1] Dept. of Computer Science, Florida State University
Tallahassee, Florida 32306
{burmester, breno}@cs.fsu.edu
[2] Dept. of Electrical and Computer Engineering, University of Maryland
College Park, MD 20742
gligor@eng.umd.edu
[3] Sch. of Computer Science, Carleton University
Ontario, K1S 5B6 Canada
kranakis@scs.carleton.ca
[4] Dept. of Electrical Engineering and Computer Science, University of California
Berkeley, CA 94720-1776
doug.tygar@gmail.com
[5] Dept. of Software and Information Systems, University of North Carolina
Charlotte, NC 28223
yzhenguncc.edu

Abstract. This paper contains the summary of a panel on authentication in constrained environments held during the Secure MADNES'05 Workshop. These were transcribed from hand-written notes.

Mike Burmester (M. B.): We are having a talk here, on the topic of *authentication on constrained environments*. The panelists could take it into any direction they find interesting. I chose this topic because it is almost impossible to achieve authentication in ad-hoc settings. (After this opening, M. B. introduces the speakers to the audience filling the room.)

Virgil Gligor (V. G.): So, I will try to be respectful to the topic. I will mention several topics that are related to the main topic.

> **First primitive:** *Authenticated encryption.* Involves 1-pass through data, requiring block cipher computation plus redundancy check. No extra work is required, only the block cipher. This is the most efficient approach. If authentication fails, then encryption fails with high probability. Note that 1-pass authenticated encryption is what you want. Alternative: Kerberos authentication (using hashes) and a second pass with block cipher. Another alternative: use block cipher and authentication (using block cipher to implement both primitives). Again, two passes over the data. The 1-pass authenticated encryption is clearly most efficient.

M. Burmester and A. Yasinsac (Eds.): MADNES 2005, LNCS 4074, pp. 186–191, 2006.
© Springer-Verlag Berlin Heidelberg 2006

Second primitive: Another form of authentication that is needed is to authenticate based on *threshold voting:* Suppose there are m neighbors in a location. One should be able to figure out if t-out-of-m vote to take some action (sensing event). Any t messages out of m authenticate in a neighborhood. How to do this? Various primitives are available: I have described a technique using *random polynomials*, where the degree should be approximately equal to the threshold. May have to use *Merkle trees*. In summary, it is possible to implement t-out-of-m authentication efficiently using one of Merkle trees, hash trees, or random polynomials.

Other form of authentication: Base station based as opposed to fully distributed. Not so interesting, and [I am] not convinced it is needed [for] more sophisticated forms of authentication.

Doug Tygar (D. T.): About base-station to node not being interesting: It hurts, I wrote a book on the topic! I do not think authentication will be a problem, at least machine-to-machine or message authentication. I believe our model is wrong. I will give an example with dual-core devices: a low-power one that runs all the time, and a high-power crypto device that works infrequently. What about tamper-resistance? I believe that trends coming in the next few years (Playstation DRM technology, trusted computing base (TCB)) will make tamper-resistance available. Important question: Will we have any anonymity in this type of environment?

Yuliang Zheng (Y. Z.): My interest is efficient cryptographic solutions to problems. I take a different approach. How keys get distributed, updated? How to design efficient/better algorithms? How to design authenticated encryption? For that, a solution is available. So, first issue: Key distribution. second issue: If we look to far ahead, we may loose perspective. Why not look at novel techniques for authentication if you want to anticipate MADNES 2010? Before then quantum devices may be available, so why not look at novel approaches?

M. B.: Take another angle. MANETS are short-term networks, but also useful as support for fixed networks. Attackers of fixed-line networks exploit particular weaknesses of the network. I wish to set-up wireless overlay networks to set-up new networks quickly and compensate on attacks on wire-based networks.

  – Issue: Key revocation. How to revoke keys? How to decide with mobile devices are not behaving as required, how to revoke them? There is scope for ad-hoc networks that are not only sensors, not short-lived, but needed for long-term survivability (using threshold-based encryption).

Evangelos Kranakis (E. K.): These guys have left nothing for me. Let me see, authentication in constrained devices. One word at a time.

  – Authentication: Verify that something is genuine. How much authentication is satisfactory to us? How much do we want? How may we afford it? Each application has its own risks, constraints, so focus on what is important.

- Constrained devices: Which part is constrained? Do we mean bandwidth limitations, resource limitations? Authentication is best for short periods of time. Authentication is short-lived.
- Environment: Assuming an ad-hoc environment, not previously thought-out. Out-of-blue authentication needed. What is involved: Servers, a wi-fi network? Environment is not well defined. Within this context, think differently, new approaches.
  - Use directional antenna in some clever way.
  - Use location in some clever way.
  - Use radio-frequency fingerprint, at least study for what aspects they may work.
  - What is possible? Measure it.

History of wired networks: As V. G. mentioned, we have experience transferring solutions from other contexts to see what can be better: Take wireless out, put wire-line back in, go back 30 years (old devices were constrained). Think differently, not just transfer solutions.

M. B.: It occurs to me, we started this workshop with V. G. with talk about how to do things in a different way. Then E. K. talked about new technologies, using non-cryptographic authentication for identification. It is an emerging technology to deal with new types of properties referring to wireless technologies.

Y. Z.: We talked about constrained devices. My watch is powered by light. Perhaps the battery issue will go away.

E. K.: I do not think battery issues will go away, it will not because of Moore's law.

Member of Audience: New technologies (TOSHIBA) increase battery capacity 3 times, then double the capacity again. Batteries improve, but many interests require power. Are we making mistakes to use crypto that is secure by going back in time 30 years?

D. T.: I do not think that we go back in time, some issues are revisited, but new issues appear. Batteries have Moore's law, everything has Moore's laws except human intelligence. Batteries could use temperature differentials, liquid fuel, could burn fuel for decades. I think these issues will be addressed. Think about the motivation: More than a third of the weight that American soldiers carry is batteries.

V. G.: Batteries are nice targets of attacks. Authentication protocols are not designed to protect against battery depletion attacks, especially public-key authentication. The problem is not really public-key if you use only once. But where to stop authentication? For example, vehicular protocols, authenticating distances. What do you mean? Feet, yards? I am talking about revealing positions. Can A reveal his position to B, then B to A? What primitives, batteries need to be used?

E. K.: There are examples of environments that are not constrained. E.g.: Internet. We try to run BGP, notoriously insecure. Cannot do anything very well. How can we be secure when we cannot do anything right?

D. T.: We cannot defend BGP. May be people will do something right next time.

V. G.: More than BGP, look at DNS. We know DNSSEC for 10 years, but it is nowhere to be seen. How to reconcile competing economic interests that are difficult to change? We know how to do Internet authentication, but other issues come in the picture. I hope in MANETs we should not box ourselves in. Back to D. T., we are interested in privacy. What is the impact of authentication on privacy? When are we for authentication? When for privacy?

D. T.: Example: My house is burglarized. I can cancel my credit cards w/o authentication. There are clear DoS issues in this picture. I want to digress. Legacy of Internet: Remember TCP/IP were experimental protocols. Do it over again, and we could make better choices. I fell negative about IPv6, its a patch instead of an evolution. In MANETs there is an opportunity not to build upon previous mistakes. What stops us from making new mistakes? Make new protocols modular so that we can address these.

Member of Audience: Regarding anonymity. Any possibility of achieving privacy-preserving authentication? Any hope?

D. T.: I do not know.

V. G.: Skeptical.

E. K.: Product of anonymity and authentication equals 1.

Y. Z.: I think devices that, for devices that are not associated with human beings, like sensors in a forest, this is not an issue. But for cell phones, this is an important issue.

Member of Audience: Some networks have hundreds of nodes. If a small number of nodes is compromised, how much damage takes place?

E. K.: A containment issue. I do not know the answer.

M. B.: Hard issue in fixed-line networks. Different, much harder in MANETs. Nearly impossible in general, but maybe local containment is achievable. Depends on the application. If it is a military application, you can use sophisticated techniques. Varies with context.

V. G.: If devices can be captured, all authentication protocols will fail. Authentication is designed to protect against man-in-the-middle attacks. If one end is compromised, what will happen to them? If adversary captures node. We will develop and deploy these technologies, introducing new vulnerabilities on old security protocols. What worries me is mismatch: That spells trouble.

D. T.: To be controversial, if [adversary] captures nodes, and re-uses them, ok. What if cloning? Need to develop cloning-proof technology.

V. G.: We have examples of protocols that can solve such problems, but they are few and far between.

D. T.: Give me 18 months. One solution: tamper-resistance. Substantial work has been achieved on this from PODC. We can learn from them. We have different applications in mind. I look at ubiquitous devices (not hundreds, but tens of thousands). Instead of looking at one sensor, pool many sensors to find answer. One question is how to design protocols to achieve security this way.

M. B.: We can handle faults, but not malicious faults. If one looks at military applications, look at the cost of attacks. Malicious faults would include cloning millions of devices, flooding. Will it happen? Who is the adversary?

V. G.: Comment about device capture: Tamper-proof not withstanding, if you have control (capture) you win (modify inputs). In World War II, spies were captured, and provided with modified information, and the Brits had poor intelligence for 2 years. This is a problem with device capture.

Member of Audience: Should remote sensing be used instead of sensors?

V. G.: Deployment has not been looked at, it is an important issue.

Y. Z.: Authentication itself is not alone the solution. Self-destruction is the answer.

V. G.: What about changing inputs? It is a fundamental problem.

Member of Audience: What about scalar authentication?

E. K.: How to mathematize this problem?

Y. Z.: Crypto technology is one of many aspects. Usability, practicability are important.

D. T.: What it means to authenticate w/ some probability? For instance, Zero-Knowledge. But not sure what that means otherwise. Once a vulnerability is discovered, authentication problems spread like wild-fire.

V. G.: People at Ericsson have looked at fairly weak authentication. It depends on the purpose. Could be that what you need if you have ways to confuse.

Member of Audience: What do you mean by weak?

V. G.: Other means of authentication: Fingerprinting, biometrics, weak for a variety of reasons. However, sufficient.

D. T.: Question for the audience: If we have a student that wants to work on this issue? What should we give him as a topic?

V. G.: Characterize your problem: Novel, formulatable, 3–4 years. E.g.: Study of emergent properties. Emergent algorithms and protocols. What is the big security problem? The big ubiquitous device is the mobile phone.

Member of Audience: People are using mobile phones as wallets, credit-cards, everything.

V. G.: Japan had an experiment with mobile phones as wallets, not very promising.

Member of Audience: If authentication is there, what about misconfiguration?

M. B.: This is like an insider attack.

V. G.: Excellent topic: Intruder attacks. We look at intrusion detection in wired networks. What to do to protect network against malicious security administrators? Right now, nothing. How to make it robust? Nothing so far.

E. K.: Some previous work exists.

Member of Audience: I think using cell phones will be secure in the future. For the moment, not feasible. But with IPv6, there will be ways to place your info somewhere in cyberspace. Currently not feasible, but with unified protocol with WLAN. IP is common protocol for all these. If we put security on IP, all these networks will be protected. Future of business is cross-protocol. I hope that in the future putting secrets on cell phones will be useful, but not now.

V. G.: Unification is good. Good research topic: Design human interfaces for security. E.g.: security interface for setting PIN/password. Did not work on my phone after switching off. I failed, and needed to look up the manual. Nice to have intuitive interface.

Member of Audience: At the moment, all in cell phones relates to 3GPP. All implementations, so it is fundamental to secure the 3GPP protocol.

V. G.: Problems: administrators fully trusted, problems with misconfiguration, bad user interfaces. It will stay like this for a while. We will get it wrong, even if protocols are good, even with tamper-resistance.

E. K.: We want to use cell phone like a wallet, but want it to be more like a pet-device–recognizing you all the time. (Human-computer authentication.) Move Ph.D. students to do some practical work.

(Panel adjourns)

Author Index

Lecture Notes in Computer Science

For information about Vols. 1–3988

please contact your bookseller or Springer

Vol. 4037: R. Gorrieri, H. Wehrheim (Eds.), Formal Methods for Open Object-Based Distributed Systems. XVII, 474 pages. 2006.

Vol. 4036: O. H. Ibarra, Z. Dang (Eds.), Developments in Language Theory. XII, 456 pages. 2006.

Vol. 4035: T. Nishita, Q. Peng, H.-P. Seidel (Eds.), Advances in Computer Graphics. XX, 771 pages. 2006.

Vol. 4034: J. Münch, M. Vierimaa (Eds.), Product-Focused Software Process Improvement. XVII, 474 pages. 2006.

Vol. 4033: B. Stiller, P. Reichl, B. Tuffin (Eds.), Performability Has its Price. X, 103 pages. 2006.

Vol. 4032: O. Etzion, T. Kuflik, A. Motro (Eds.), Next Generation Information Technologies and Systems. XIII, 365 pages. 2006.

Vol. 4031: M. Ali, R. Dapoigny (Eds.), Innovations in Applied Artificial Intelligence. XXIII, 1353 pages. 2006. (Sublibrary LNAI).

Vol. 4029: L. Rutkowski, R. Tadeusiewicz, L.A. Zadeh, J. Zurada (Eds.), Artificial Intelligence and Soft Computing – ICAISC 2006. XXI, 1235 pages. 2006. (Sublibrary LNAI).

Vol. 4027: H.L. Larsen, G. Pasi, D. Ortiz-Arroyo, T. Andreasen, H. Christiansen (Eds.), Flexible Query Answering Systems. XVIII, 714 pages. 2006. (Sublibrary LNAI).

Vol. 4026: P.B. Gibbons, T. Abdelzaher, J. Aspnes, R. Rao (Eds.), Distributed Computing in Sensor Systems. XIV, 566 pages. 2006.

Vol. 4025: F. Eliassen, A. Montresor (Eds.), Distributed Applications and Interoperable Systems. XI, 355 pages. 2006.

Vol. 4024: S. Donatelli, P. S. Thiagarajan (Eds.), Petri Nets and Other Models of Concurrency - ICATPN 2006. XI, 441 pages. 2006.

Vol. 4021: E. André, L. Dybkjær, W. Minker, H. Neumann, M. Weber (Eds.), Perception and Interactive Technologies. XI, 217 pages. 2006. (Sublibrary LNAI).

Vol. 4020: A. Bredenfeld, A. Jacoff, I. Noda, Y. Takahashi (Eds.), RoboCup 2005: Robot Soccer World Cup IX. XVII, 727 pages. 2006. (Sublibrary LNAI).

Vol. 4019: M. Johnson, V. Vene (Eds.), Algebraic Methodology and Software Technology. XI, 389 pages. 2006.

Vol. 4018: V. Wade, H. Ashman, B. Smyth (Eds.), Adaptive Hypermedia and Adaptive Web-Based Systems. XVI, 474 pages. 2006.

Vol. 4017: S. Vassiliadis, S. Wong, T.D. Hämäläinen (Eds.), Embedded Computer Systems: Architectures, Modeling, and Simulation. XV, 492 pages. 2006.

Vol. 4016: J.X. Yu, M. Kitsuregawa, H.V. Leong (Eds.), Advances in Web-Age Information Management. XVII, 606 pages. 2006.

Vol. 4014: T. Uustalu (Ed.), Mathematics of Program Construction. X, 455 pages. 2006.

Vol. 4013: L. Lamontagne, M. Marchand (Eds.), Advances in Artificial Intelligence. XIII, 564 pages. 2006. (Sublibrary LNAI).

Vol. 4012: T. Washio, A. Sakurai, K. Nakajima, H. Takeda, S. Tojo, M. Yokoo (Eds.), New Frontiers in Artificial Intelligence. XIII, 484 pages. 2006. (Sublibrary LNAI).

Vol. 4011: Y. Sure, J. Domingue (Eds.), The Semantic Web: Research and Applications. XIX, 726 pages. 2006.

Vol. 4010: S. Dunne, B. Stoddart (Eds.), Unifying Theories of Programming. VIII, 257 pages. 2006.

Vol. 4009: M. Lewenstein, G. Valiente (Eds.), Combinatorial Pattern Matching. XII, 414 pages. 2006.

Vol. 4008: J.C. Augusto, C.D. Nugent (Eds.), Designing Smart Homes. XI, 183 pages. 2006. (Sublibrary LNAI).

Vol. 4007: C. Àlvarez, M. Serna (Eds.), Experimental Algorithms. XI, 329 pages. 2006.

Vol. 4006: L.M. Pinho, M. González Harbour (Eds.), Reliable Software Technologies – Ada-Europe 2006. XII, 241 pages. 2006.

Vol. 4005: G. Lugosi, H.U. Simon (Eds.), Learning Theory. XI, 656 pages. 2006. (Sublibrary LNAI).

Vol. 4004: S. Vaudenay (Ed.), Advances in Cryptology - EUROCRYPT 2006. XIV, 613 pages. 2006.

Vol. 4003: Y. Koucheryavy, J. Harju, V.B. Iversen (Eds.), Next Generation Teletraffic and Wired/Wireless Advanced Networking. XVI, 582 pages. 2006.

Vol. 4001: E. Dubois, K. Pohl (Eds.), Advanced Information Systems Engineering. XVI, 560 pages. 2006.

Vol. 3999: C. Kop, G. Fliedl, H.C. Mayr, E. Métais (Eds.), Natural Language Processing and Information Systems. XIII, 227 pages. 2006.

Vol. 3998: T. Calamoneri, I. Finocchi, G.F. Italiano (Eds.), Algorithms and Complexity. XII, 394 pages. 2006.

Vol. 3997: W. Grieskamp, C. Weise (Eds.), Formal Approaches to Software Testing. XII, 219 pages. 2006.

Vol. 3996: A. Keller, J.-P. Martin-Flatin (Eds.), Self-Managed Networks, Systems, and Services. X, 185 pages. 2006.

Vol. 3995: G. Müller (Ed.), Emerging Trends in Information and Communication Security. XX, 524 pages. 2006.

Vol. 3994: V.N. Alexandrov, G.D. van Albada, P.M.A. Sloot, J. Dongarra, Computational Science – ICCS 2006, Part IV. XXXV, 1096 pages. 2006.

Vol. 3993: V.N. Alexandrov, G.D. van Albada, P.M.A. Sloot, J. Dongarra, Computational Science – ICCS 2006, Part III. XXXVI, 1136 pages. 2006.

Vol. 3992: V.N. Alexandrov, G.D. van Albada, P.M.A. Sloot, J. Dongarra, Computational Science – ICCS 2006, Part II. XXXV, 1122 pages. 2006.

Vol. 3991: V.N. Alexandrov, G.D. van Albada, P.M.A. Sloot, J. Dongarra, Computational Science – ICCS 2006, Part I. LXXXI, 1096 pages. 2006.

Vol. 3990: J. C. Beck, B.M. Smith (Eds.), Integration of AI and OR Techniques in Constraint Programming for Combinatorial Optimization Problems. X, 301 pages. 2006.

Vol. 3989: J. Zhou, M. Yung, F. Bao, Applied Cryptography and Network Security. XIV, 488 pages. 2006.